Praise for *Grass-Fed Beef for a Post-Pand*

"Grazing animals are a natural part of the land. When grazing is done right, it will improve soil health and regenerate the land. Cattle have been criticized for occupying too much land, but people forget that 20 percent of the habitable land on Earth cannot be used for crops. The only way to raise food on this land is grazing animals. The land is either too hilly or arid for crops. In this book, you will learn how grazing and regenerative agriculture is a win-win for both producing food and the environment."

—**TEMPLE GRANDIN**, author of *Animals Make Us Human*; professor, animal science, Colorado State University

"The next time someone argues that cows are disastrous for the planet, hand them a copy of *Grass-Fed Beef for a Post-Pandemic World*. Equal parts manifesto and how-to guide, Shinn and Pledger will show you that the solution to our human and planetary health crisis begins with a cow eating grass and ends with the most delicious steak you've ever had."

—**DAN BARBER**, author of *The Third Plate*

"This comprehensive and engaging account offers a path forward from industrial to regenerative agricultural practices, one that is urgently needed in the face of diminishing availability and increasing costs of the fossil fuels largely responsible for a precipitously warming global climate. This book is a must-read for people worldwide interested in how managed grazing can enhance the health of soil, plants, domesticated and wild animals, and humans and can help cool a warming planet as increasing temperatures make large swathes of the globe, including many parts of the United States, uninhabitable during the next fifty years."

—**FRED PROVENZA**, professor emeritus, Department of Wildland Resources, Utah State University; author of *Nourishment*

"Lynne Pledger and Ridge Shinn have created a readable, usable guide to grass-fed beef—full of both the hows and whys. An essential addition to the libraries of everyone involved in the raising and selling of beef."

—NICOLETTE HAHN NIMAN, author of *Defending Beef*

"Whether you eat beef or not, this book reveals what everyone needs to know—why grass-fed beef is better not just for the health of cows but for people and the planet as well."

—ANNE BIKLÉ, coauthor of *What Your Food Ate*

"Challenging the entrenched wisdom that cows are bad for us and the environment, *Grass-Fed Beef for a Post-Pandemic World* offers a well-seasoned perspective that the real problem is how we raise them."

— DAVID R. MONTGOMERY, author of *Dirt* and *Growing a Revolution*

"Anchored in the science, history, and first-hand practice of regenerative grazing, Ridge Shinn and Lynne Pledger make a deeply informed and unromanticized case for incorporating the 'work with nature' principles of nineteenth century agriculture to, among other things, restore soils, promote rural economies, mitigate climate disruption, and support overall well-being—system-wide and at scale."

—JOCK HERRON, design critic focused on food systems and health, Harvard Graduate School of Design

"*Grass-Fed Beef for a Post-Pandemic World* is a well-researched and timely contribution to a much-needed conversation about what we eat and where it comes from. Fascinating for anyone interested in finding ways they can personally help mitigate climate change and 'eat better' for the planet, for animal welfare, and for their own health. Essential reading for farmers wanting to restore their land, feel good about what they do, and turn a profit."

—LIBBY HENSON, codirector and cofounder, Grassroots Systems

"As we increasingly recognize the vital role of regenerative grazing in human, ecological, and planetary health, the question invariably asked is, how quickly can it be scaled up? Informed by decades of industry experience and market success, the authors of *Grass-Fed Beef for a Post-Pandemic World* lay out a brilliant strategy for transforming beef production from a conventional, extractive, fossil fuel–intensive model to an approach that heals degraded soil, improves wildlife habitat, rejuvenates rural economies, and sequesters carbon. Best of all, their vision is adaptable to all regions."

—**KARL THIDEMANN**, cofounder, Soil4Climate

"This book gives me hope. In it, soil and climate heroes Ridge Shinn and Lynne Pledger bring us a giant step closer to the regenerative future. Buy it. Absorb it. Cherish it. Share it."

—**SETH J. ITZKAN**, cofounder, Soil4Climate

"This timely book by Ridge Shinn and Lynne Pledger masterfully covers how a transition to regenerative grazing can restore ecosystem function to deliver vital ecosystem services to provide the ecological and economic resilience required for a secure and healthy food system base. They outline the steps necessary to transition from the current food system organization to systems that facilitate decision making on the land and educate the population on the benefits of managing to restore ecosystems."

—**RICHARD TEAGUE**, professor,
Texas A&M AgriLife Research

Grass-Fed Beef for a Post-Pandemic World

How Regenerative Grazing Can Restore Soils and Stabilize the Climate

RIDGE SHINN and **LYNNE PLEDGER**

Foreword by **GABE BROWN**

Chelsea Green Publishing
White River Junction, Vermont
London, UK

Project Manager: Angela Boyle
Acquiring Editor: Benjamin Watson
Developmental Editor: Ben Trollinger
Copy Editor: Diane Durrett
Proofreader: Laura Jorstad
Indexer: Linda Hallinger
Designer: Melissa Jacobson
Page Layout: Abrah Griggs

Printed in Canada.
First printing October 2022.
10 9 8 7 6 5 4 3 2 1   22 23 24 25 26

ISBN 9781645021247 (paperback) | ISBN 9781645021254 (ebook) | ISBN 9781645021261 (audio book)

Library of Congress Cataloging-in-Publication Data is available upon request.

Chelsea Green Publishing
85 North Main Street, Suite 120
White River Junction, Vermont USA

Somerset House
London, UK

www.chelseagreen.com

For Jacob, Jasper, and Mariah

Contents

Foreword

I t seems that every day we wake up to a new crisis, from the COVID-19 pandemic and high fuel prices to empty store shelves and war in Ukraine. Our world is in a constant state of change. While many cry doomsday, some, like Ridge Shinn and Lynne Pledger, know that the secret to resiliency in a changing world can be found through healthy soil, created with the help of grazing ruminants.

Relying on a lifetime of experience with cattle and resource conservation, Ridge and Lynne take us on a journey through production agriculture in the United States. They explain how farmers and ranchers went from grazing livestock on pasture to confining cattle in grassless pens where grain and by-products are brought to them—grain and by-products that are grown with synthetics (fertilizers, pesticides, and fungicides) and the heavy use of fossil fuels. These practices, in no small part, have led to the degradation of our soils, our waters, and, at least to some degree, our health.

This has led many to vilify animals, cows in particular, and falsely claim that they are largely responsible for climate change. This book helps to set the record straight: This new method of grazing cattle has a net climate benefit. As Ridge and Lynne so truthfully state, "Regenerative grazing will change the way our society thinks about beef, because the grazing itself is as significant as the meat."

The act of grazing as significant as the meat itself? Yes! Few are aware of just how important it is for plants to be grazed. For eons, the simple act of an herbivore biting off part of a plant has led to what Dr. Allen Williams refers to as "positive compounding effects."

The herbivore, such as a cow, bites a plant; the plant releases root exudates (carbon compounds); fungi and bacteria feed on the exudates; protozoa and nematodes feed on the fungi and bacteria, excreting excess nitrogen, which is then used by the plants for growth. Microorganisms also help to bind sand, silt, and clay particles together into aggregates, and those aggregates allow water to infiltrate the soil instead of flowing off over the surface. Infiltrated water absorbed into the carbon in the soil helps make farmland resilient to drought.

The positive compounding effects generated by grazing animals go on and on. Ridge and Lynne take a deep dive into this soil health, plant health, animal health, and human health connection. It is this connection that has the ability to bring us together. No matter where your interest lies—be it climate change, clean water, profitability for farmers and ranchers, the revitalization of our rural communities, or improved human health—regenerative agriculture, driven by adaptively grazed cattle, can address many of the issues we face as a society. We can come together, finding common ground for common good.

I believe that we must continue to educate ourselves, no matter what profession we are in. To keep honing our skills, we must learn from those with experience. In *Grass-Fed Beef for a Post-Pandemic World*, Ridge and Lynne share their lifetimes of experience. In fact, every chapter highlights stories of people, places, and predicaments that Ridge encountered over the years as his understanding of grass-fed beef production developed.

I learned from this book and am confident that you will, too.

—*Gabe Brown*
June 2022

Preface

Sometimes I say that I learned to farm in the nineteenth century.

Two years out of college, I quit my job as an environmental advocate in Washington, DC, to become a farmer at a New England living history museum. Driving north to Old Sturbridge Village, I was more than happy to leave the vagaries of Capitol Hill politics behind. From early childhood, I have been interested in farming above all else. My new job would allow me to research agricultural practices of the early 1800s and work outdoors through the seasons, using old-style tools and implements in the museum's fields and barns.

At the living history site, I met Lynne Pledger, who was already working there in a position parallel to mine, researching and demonstrating farm tasks performed by rural women in the early nineteenth century, including milking cows and making butter and cheese.

With jobs that immersed us in cow-centric daily routines, Lynne and I came to realize that farm families in that era (actually right up until World War II) managed their cattle in ways that worked with nature, rather than against it.

- Cows were bred to calve in the spring when the grass started growing.
- Cattle were raised and fattened on grass and forage, with hay in winter.
- Corn was raised for human consumption as cornmeal— not for cattle.
- Cattle deposited their manure and urine all over the pasture, fertilizing the soil.

Of course, not all farm practices from the good old days were good. For example, farmers plowed their fields for annual crops. Now we know that plowing allows soil carbon to oxidize, releasing it to the atmosphere, whereas perennial pastures for livestock keep the soil protected by vegetation. But considering the global harm from today's industrial agriculture, in many ways the "old-fashioned" approach looked better to us than what replaced it: chemical fertilizers and biocides and the confinement of livestock to CAFOs (confined animal feeding operations).

In 1973, Lynne and I married and bought thirty-four acres in rural Massachusetts, where we planted organic gardens and fruit trees, kept honeybees, tapped maples for syrup, and raised livestock. We had two children and now have three grandchildren. Then, after forty-five years, we divorced. But throughout those decades and continuing right up to the present, we have worked together on farming and sustainability issues.

Even in the early days of our marriage, the scope of our agricultural interests reached beyond our homestead. We had barely settled into our rural community in the early 1970s when we became painfully aware of the decline of dairy farming all around us. Every time I heard that another local farm was going out of business, it felt like a death in the family. I realized then that developing a productive homestead would never be enough for me; I wanted nothing less than to live in a thriving agricultural economy—though I did not foresee how agriculture in the Northeast could regain its former glory or how I might contribute to that outcome.

Then a project beckoned that grew out of work we had begun at Old Sturbridge Village, when we were searching for authentic breeds of livestock to populate the fields and barnyards of the living history museum. At that time we had started an organization to identify and promote endangered breeds of livestock in the United States: cattle, sheep, goats, pigs, horses, and poultry. In 1977, a small group of us renewed our commitment to that goal and I became the executive director of the American Minor Breeds Conservancy, while Lynne published the organization's newsletter. In 1983 our efforts were recognized in an article in *Smithsonian* magazine, "Farm Animals: Saving America's Rare Breeds." In 1985 I gave up the leadership role, and the organization moved from our kitchen table to Pittsboro, North Carolina, where it continues as The Livestock Conservancy.

For a time, Lynne and I raised heritage breeds of pigs. Our farm was certified organic and our pigs had the run of the fields and woods. But pigs are not ruminants; even pastured pigs need grain, and organic grain is expensive. With other organic livestock farmers in our area, we would put in a group order for grain, and every month the tractor trailer pulled into our yard and the other farmers would come by and pick up their share. Raising pigs began to feel like a materials-handling exercise. Eventually, the expense and logistics of buying organic grain made raising pigs unappealing. Raising cattle was more agreeable to me. Because cattle are ruminants, they can and should be raised on grass and forage only—no grain ever. I became committed to 100% grass-fed beef, and raised and marketed Devon cattle instead of pigs.

During the first decade of the millennium, with a succession of business partners in various start-ups, I gained experience working with other grass-fed beef producers and developing markets and distribution systems in the Northeast. In 2010 I was featured in an article in *TIME* magazine that focused on the potential of grass-fed beef production to combat climate change; a photo caption in the article dubbed me a "carbon cowboy." By 2013 I had traveled all over this country and overseas as a grass-fed beef consultant.

Lynne and I began to think about establishing an environmentally beneficial and economically feasible model of widespread grass-fed beef production in the Northeast. The potential for grass-fed beef to revive family farming in our own region was compelling. In 2015, with advice and support from family, friends, and a volunteer business development committee, we organized Grazier LLC, and I began actively seeking Northeast farmers interested in raising cattle for a grass-fed beef program.

In 2017 our company went into the marketplace doing business as Big Picture Beef. Lynne chose the name to reflect the interdependence of everything on the planet: microbes, soil, water, plants, animals, and of course, people—the big picture. From the beginning, we have been committed to producing healthy 100% grass-fed beef in a way that benefits the environment, and marketing the beef in a way that provides a fair return to the farmers who raise and fatten the cattle.

In recent years global news has been dire. The COVID-19 pandemic highlighted the need for local and regional supplies of meat; we can no longer rely on beef processed thousands of miles away from our homes. And a February 2022 report from the United Nations says that irreversible climate change is coming sooner than scientists had thought; as a nation we can no longer support corn-fed cattle production, which contributes significantly to greenhouse gas emissions. Instead, each region can raise its own healthy beef cattle, using humane husbandry and pasture management that rejuvenates farmland and combats climate change by storing carbon in the ground. In these pages we describe how this can be done.

—*Ridge Shinn*

Introduction

Grass will act as the great balance wheel and stabilizer to prevent gluts of other crops—to save soil from destruction—to build up a reserve of nutrients and moisture in the soil, ready for any future emergency, to create a prosperous livestock industry, and finally to contribute to the health of our people through better nutrition.

—HENRY A. WALLACE[1]

Regenerative grazing will change the way our society thinks about beef, because the grazing itself is as significant as the meat.

In recent years, people have recognized that the conventional system of confining cattle in feedlots and raising corn to fatten them is harmful, inhumane, and unsustainable. Industrial corn production degrades farm fields, and the runoff of fertilizers and biocides pollutes streams and rivers. Corn-fed cattle suffer from a variety of diseases in feedlots, and their meat poses health threats to consumers. Worst of all, conventional beef production is a significant source of climate emissions.

But we mustn't take beef off the table. The problems associated with feedlot beef—including methane burps—stem from the way the animals are raised. In contrast, regenerative grazing, which is the best methodology for producing 100% grass-fed beef, has no resemblance to the conventional mode of beef production. This methodology is a low-tech approach that works with existing natural systems (such as photosynthesis and nutrient cycling), and can be part of the solution to some of our most confounding problems: depleted soils, droughts and floods, nutrient-deficient food, food shortages, and even climate change. Ironically, some entrenched but misguided beliefs about beef are working against the widespread adoption of this environmentally beneficial method of raising cattle.

The principles of regenerative grazing were developed by pioneering agriculturalists in the twentieth century, and those principles have now been applied successfully in many parts of the world. In this book we reference regenerative farms and ranches all over the United States as well as in Canada and Mexico. In addition to producing beef that is a healthy source of protein (see chapter 4), regenerative grazing and pasture management rejuvenates degraded farmland and increases the amount of carbon stored in the soil (chapter 3). It also addresses the needs of cattle as ruminants and herding animals (chapter 5).

This new approach is especially relevant since the pandemic; there have been long-standing concerns about conventional beef production, but the outbreak of COVID-19 revealed some new risks posed by our concentrated meat industry—risks that regional grass-fed beef production can address.

Shocks to the Food System

In the spring of 2020, pandemic-related labor shortages and shipping delays caused food distribution woes that were front-page stories, illustrated by photos of spoiled fruits and vegetables. In a world with hunger on the rise, the public saw mountains of wasted food.[2] Perhaps some of it went to compost facilities, but much of it was surely destined for landfills or incinerators. In addition, a number of giant meatpacking facilities closed abruptly when assembly-line workers contracted COVID-19. When a facility that normally slaughters thousands of animals in a single day closes its doors, there is no way to handle the scale of the backlog. The supply chain is designed for efficiency and is not prepared to withstand major shocks. Even more distressing than the photos of spoiled produce were photos of live animals culled for slaughter, their carcasses to be discarded.

Many consumers, concerned about a meat shortage, headed for grocery stores and filled their carts. As days went by and meat cases were emptying, prices skyrocketed, but shoppers continued to buy. Because some of the corporations that own processing plants also control the whole supply chain and sell directly to the stores, this was a boom time for them. But the high revenues did not reach back to the farmers and ranchers who sell their cattle to the big corporations—a long-standing disparity that has cattle producers angry about their small share of the dollars that consumers spend on beef.[3]

Jennifer Clapp, a political economist who studies food, posed both the problem and the solution in the *New York Times*:

> We need to rejuvenate local and regional food systems to reduce the vulnerabilities that come with being too reliant on imported and corporate dominated foods. This doesn't mean cutting off all trade or abolishing all packaged foods, but it does mean building diversity, decent livelihoods for workers, and opportunities for small and medium-scale enterprises to flourish in shorter, more sustainable food supply chains that are closer to home.[4]

Then, in June 2021, just as COVID-19 cases ebbed across the United States, a ransomware attack shut down production in nine JBS Foods meat-processing plants for a couple of days. JBS processes 20 percent of the US beef supply, so even this brief interruption brought cattle prices down (hurting farmers and ranchers) and brought processing profits up.[5]

Developing sources of beef in each region of the country would be a key step in building regional resilience against shocks such as pandemics, ransomware attacks, natural disasters, terrorism, or wars. These threats are no longer hypothetical.

Everybody Eats

Long before the pandemic forced meat-processing plants in the United States to close, voices from the fields of food and agriculture were calling for better ways to produce healthy beef and get it from farm to table. But it took the global catastrophe to prompt even broader public discussions on the perils of our nation's highly concentrated food system, including the well-being of the people who raise the animals or package the beef.

For at least a couple of decades, increasing numbers of consumers have rejected the feedlot model, preferring to spend their food dollars on sustainably produced beef from cattle that graze in pastures for their entire lives, consuming no corn or grain at all. During the same period, 100% grass-fed beef producers around the country have been honing their skills, sharing experiences, measuring increases of carbon in their soil, and looking for markets for their beef. Now they are finding customers.

By the time the closure of big meat plants in 2020 had called the centralized US food system into question, in the Northeast we had developed a new approach to producing 100% grass-fed beef, raised by regenerative methods, and had gone into the marketplace in 2017 doing business as Big Picture Beef.

We work with farmers* throughout our region to produce beef for wholesale buyers in all six New England states (Maine, Vermont, New Hampshire, Massachusetts, Connecticut, and Rhode Island) plus New York, New Jersey, Pennsylvania, Delaware, Maryland, and West Virginia: Northeast grass-fed beef for Northeast consumers. Our strategy for developing a consistent regional supply of 100% grass-fed beef has three bedrock elements:

1. Participating farmers in the region adopt the specific practices for healthy soil and healthy cattle that are collectively called regenerative grazing, which we explain in detail in the coming chapters.
2. The final phase of production—finishing—is managed in pastures by graziers experienced in fattening cattle on grass and forage only, with no grain.
3. The whole system is sustainable economically, as well as ecologically. True sustainability means that everybody eats: not just the cattle and the consumer, but also soil microbes, pasture plants, cattle producers, facility workers, and vendors.

Given that our Northeast farms are relatively small, in order to provide enough volume for wholesale markets, Big Picture Beef partners with multiple farmers in an area to raise the young stock according to our standards and protocol.† Then we aggregate cattle from a number of those cow-calf farms on one of our "finishing" farms, where an experienced grazier fattens

---

* Which term to use, *farmer* or *rancher*, is largely a matter of where you live. West of the Mississippi, people who raise cattle are ranchers; in New England, people who raise cattle are farmers. In this book, we generally refer to them as farmers if they live in the East and as ranchers if they live in the West—or simply as "producers."

† Among other things, our Big Picture Beef protocol requires that cattle receive no antibiotics, growth hormones, or grain, and that the fields and pastures are free of biocides.

the large herd for market by moving them frequently to a new paddock of fresh grass as needed, typically multiple times per day. This skilled finish ensures that the meat will be tasty.

This strategy can be applied in other regions, as well. While the Northeast is fortunate to have good soil, abundant rainfall, and major urban markets close at hand, our basic system is adaptable to varying climates and circumstances. Regenerative grazing has been practiced in areas with as little as ten inches of rainfall a year.[6] And based on a 2016 survey, producers of grass-fed beef are already well distributed throughout the country. In 2017 there were 3,900, up from 100 in 1998. (A large number of them produced less than fifty head each year, and mostly sell directly to consumers.)[7]

There is enough grassland in the United States to finish all the cattle currently fattening in feedlots on grain.[8] The states now devoted to raising crops to feed livestock could be integrating cattle into cropping systems to great economic and ecological benefit, as some of the farmer/ranchers we reference are now doing. Both the grassland and the expertise are available for a tremendous expansion of grass-fed beef production.

A Cascade of Benefits

Regenerative grazing starts a cascade of environmental benefits. To activate this phenomenon, grass-fed beef producers refrain from conventional farming practices that degrade farmland, and instead employ grazing and pasture management strategies that are far more productive in the long term. Each farm or ranch realizes ecological gains from stewardship of the farm's soil, which typically results in net profitability.

Unlike conventional rotational grazing, in which the cattle are rotated through paddocks at set intervals, the key concept of regenerative grazing is that the grazier moves the herd into a new plot only after it has had sufficient time to regrow completely before being grazed again.

The rejuvenation of the pasture takes place below ground as well as above. In fact, the most critical players in this natural process are not the farmers and ranchers, but soil microorganisms, notably fungi and bacteria, that are essential to soil health and structure. (These microorganisms have been referred to as "the micro herd.") The agricultural practices that we describe in this book foster these helpful microbes. We will cite reports and studies from various sections of the United States that document the ecological advantages of working *with* natural soil systems instead of against them.

Raising cattle by regenerative grazing is part of the larger regenerative agriculture movement, which aims to improve the ecosystems in which the food is grown. This approach to raising cattle is the polar opposite of the conventional beef production model that has led many to believe that raising cattle is bad for the environment and consuming red meat is unhealthy. Some people call the regenerative approach "beyond sustainable" because the farmland actually improves in multiple, measurable ways.

But this isn't about sacrificing profit to do the right thing for the environment. Soil health is inseparable from profitability because this method increases the productivity of the land with fewer costly inputs like fertilizers, pesticides, and herbicides. Improving soil health improves profit margins.[9] Research ecologist Richard Teague from Texas A&M University recalls his impressions on his first visits to farms doing regenerative grazing in the southeastern United States:

> There would be much more grass even though they had twice as many animals as their neighbors. That is because they were producing more grass by managing in a way that allowed recovery of the plant leaves and roots after grazing, and this benefited the soil infiltration rates and fertility.[10]

If beef producers over a broad geographic area adopt this approach and the beef is distributed widely via wholesale buyers—stores, restaurants, and institutions—a whole region can achieve maximum health and environmental benefits. And developing appropriate infrastructure for slaughter and processing would boost rural economies and save food miles, as well.

Save the Grasslands

Before humans evolved, our planet was home to huge herds of ruminants (buffalo, sheep, deer, antelopes, and others). Predators, nipping at the hooves of stragglers, kept those herds moving slowly across the world's rangelands, allowing the ruminants just enough time to eat the high-energy tops of the plants, trample the rest, and leave manure to boost regrowth on the land they left behind. But manure was not the ruminants' only contribution to fertility.

Well-Managed Grazing—What to Call It?

The environmental benefits of well-managed grazing are well established, and there is broad agreement on the best grazing and pasture management practices: simply put, a multi-paddock rotation with short grazing periods and long periods of regrowth that allow each paddock to fully rejuvenate before it is grazed again. But what should we call the whole approach? Certainly, an agreed-upon term would be helpful for the consumer.

Many names have been tried and several are still in use. By the time this book is in print there may be more.

The most precise names are lengthy and hard to remember. In this category we think *adaptive multi-paddock grazing* (AMP) is clearer than an older term, *management-intensive grazing* (MiG). But neither is memorable or self-explanatory. *Holistic planned grazing* gives well-deserved recognition to the work of Allan Savory, pioneering ecologist and livestock producer—but it doesn't roll off the tongue, either.

Shorter names tend to be vague and associated with familiar practices that are not as transformative as what we want to describe. *Rotational grazing* or simply *planned grazing* have this drawback.

So we use several terms. In presentations where the audience is made up of agriculturalists or ecologists, we use the term AMP—adaptive multi-paddock grazing. *Adaptive* means that the grazier responds to changing conditions regarding when and where to move the herd. This term is used in peer-reviewed journals. But we also need a term to use with general audiences.

Throughout these pages, you'll see *regenerative grazing* most often, but it is not entirely self-explanatory. So we often start discussions with *well-managed grazing* and elaborate from there. We use *managed* over *planned* because the latter implies something theoretical that might never happen, whereas *managed* conveys activity on the ground. And adding *well* to *managed* implies success.

By grazing the grassland plants as they moved along, these animals were continuously triggering a series of responses from organisms beneath the sod that were critical to the maintenance and renewal of vast areas such as the Great Plains in the United States. In chapter 3 we explain why ruminant grazers are still keystone species in terms of their importance to grassland ecosystems.

The remarkable depth of topsoil on the Great Plains was the result of the synergy of diverse species playing specific roles: not only the American buffalo but also their predators, the plants they grazed, and the abundant soil microbes, earthworms, dung beetles, and other organisms. Global warming was not a problem because the ecosystem was in balance. For example, the buffalo would have emitted some methane, but the methanotrophic bacteria that were on the scene at ground level—close to the mouths of grazing bovines—would have *oxidized* some of this gas (that is, taken electrons from it) so that it would not reach the atmosphere. These bacteria perform the same function in our pastures today—but not in feedlots.[11]

Now, with large human populations and their disruptive and destructive activities, more than ever we need grasslands to conduct photosynthesis and store carbon and water. Without grazing, grass grows high, desiccates, and oxidizes, contributing to climate change. Even if nobody ate meat, we would still need ruminants to graze because of their ecological functions. Rangeland ecosystems occupy nearly half of the Earth's land area and many are degrading primarily due to inappropriate agricultural practices, which we describe in chapter 2.[12]

Today, cattle that are managed regeneratively provide the equivalent effect of the buffalo, whose numbers have been reduced from an estimated twenty-five to thirty million in North America in the sixteenth century to only thousands worldwide today.[13] Now, instead of predator populations prompting the herds to keep moving, a small but growing number of skilled agrarians manage cattle in a way that mimics the movement of those wild ruminants of an earlier era.

High Stakes

World hunger, after steadily declining for a decade, is on the rise, affecting 9.9 percent of people globally.[14] Allowing healthy soil systems to function—or not—can be the difference between abundance and scarcity. Regenerative grazing can address world hunger by increasing the yield from farm fields. While some regenerative practices can be applied to raising vegetables without ruminant livestock, it has been demonstrated that grazing cattle on cropland dramatically increases the yield of food that can be grown on that acreage.[15]

Rangelands that are not suitable for producing food for people directly (because of topography or type of plant cover) can be used for nutrition if the plants that grow there are consumed by ruminant animals. Cattle, with a digestive system that includes a rumen for fermenting grass, can convert the nutrients in cellulose (the cell walls of plants) into a meat that is an excellent food for humans.

Wherever beef producers practice regenerative grazing, they are working with natural systems that predate agriculture. Below is a partial list of eco-services that are fostered by the regenerative grazing practices we describe in this book:

carbon sequestration: offsetting greenhouse gas emissions
water infiltration and retention: protection against droughts and floods
nutrient cycling: the transport of nutrients from soil and back to soil
soil formation and fertility: deeper topsoil and reversal of desertification
biodiversity of plant and animal species: abundance above and below
    the soil line
wildlife habitat: homes for diverse animal species, large and small,
    predators and prey
ecosystem stability and resilience: withstanding climate change and
    weather events

We can no longer take any of these ecoservices for granted. They are at stake every day as people in our society make decisions: about what foods to buy, what agricultural practices to adopt, or which food or farming policies to put in place. While some of the topics in the book may be of particular interest to farmers and ranchers, most of the information is as relevant to concerned consumers and policy makers as it is to beef producers. This is not as much of a *how-to* book as a *why-to*, *can-do*, and *must-do*:

Part 1 (chapters 1 through 5) explains *why* it's important for US farmers and ranchers to replace certain grazing and pasture management practices (in use since World War II) with regenerative management. We describe how

Holistic Management in Saskatchewan

It was spring in Saskatchewan, and I was visiting some of the early adopters of regenerative grazing management. Brady Wobeser was touring me around his ranch so I could look at the cattle. As we walked through one of the paddocks, I was alarmed to spot a coyote close by, not far from cows and newborn calves.

Brady was unperturbed.

I had been at ranches where the distant howl of a coyote had sent the ranch hands jumping into four-wheelers with their guns to find and kill the animal. "Aren't you concerned that the coyote will get a calf?" I asked.

Brady explained that they bred their cattle to calve in May, when the weather is mild and the grass is growing—which is also when wild animals are having their young: squirrels, rabbits, and deer. These wild babies are easy pickings for predators. So, in May, coyotes aren't hungry enough to face a mother cow who would use her formidable bulk to protect her young one.

the updated method can restore soil health, provide nutrient-dense food, increase carbon sequestration, and more. Chapter 1 looks at the possibility of grass-fed beef reviving rural economies all over the country and considers the growing consumer demand for this product and the opportunities that family farmers have for getting a fair return for producing grass-fed cattle.

Chapters 2 through 5 provide details and documentation about the benefits of regenerative grazing and grass-fed beef from the fields of soil science, ecology, health and nutrition, and animal welfare.

Part 2 (chapters 6 and 7) presents practical ways that farmers and ranchers *can* achieve success with a grass-fed operation in a variety of climates and without a huge capital investment. Because regenerative grazing does not

Of course. The Wobeser family are adherents of Holistic Management, a framework for decision making, originally developed by Allan Savory, that considers the needs of animals, the ecosystem, and the community—everything—as a whole. A key idea is that farmers and ranchers should be guided by nature's systems whenever possible. Regarding the threat of coyotes, timing the birth of livestock to be in sync with wild animal births made everything easier.

The Holistic Management system helps farmers and ranchers manage agricultural resources to reap ecological, economic, and social benefits. The Wobeser family has applied these principles so successfully in their decision making that in 1999 Brady's parents, Dennis and Jean, were given the Emerald Award, Canada's major environmental honor, for their outstanding achievements on their ranch. The press release that accompanied the award described their low-input, nature-based operation as "truly a healthy and vibrant ecosystem."

—*Ridge*

require ownership of a vast land base and expensive equipment, it offers new opportunities for people from a variety of backgrounds to participate in our food system as owners or operators of beef cattle operations.

Chapter 6 outlines a strategy for family farms to meet wholesale buyers' requirements for volume and consistent quality. It also outlines a husbandry protocol and grazing fundamentals.

Chapter 7 provides recommendations for grass-fed beef producers to increase net profitability.

Part 3 (chapters 8 and 9) presents challenges that our country *must* overcome if we are to realize the benefits—health, environmental, and economic—of grass-fed beef production.

Chapter 8 outlines slaughter and processing challenges. In the United States the beef supply chain is controlled by a handful of multinational companies. While this can be a practical problem for grass-fed beef producers seeking affordable processing, it is also a societal problem because of unacceptable conditions for workers and unfair remuneration for farmers and ranchers. Furthermore, having the nation's meat supply highly concentrated leaves us all vulnerable in the face of any significant crisis.

Chapter 9 outlines additional challenges, including subsidies for corn, which prop up the feedlot model; legal-but-misleading labeling of the grass-fed beef sold in stores; and the growing movement to replace real meat with laboratory meat. We conclude with a call for action on grazing as an urgently needed climate strategy.

Two major problems that this book addresses—combating climate change and replacing the harmful practices of conventional beef production—are closely related. Both challenges call for updating agricultural practices. By adopting regenerative grazing methods, some farmers and ranchers have already increased both soil fertility and carbon storage.

These changes have not come about by miracles; in this book we describe the specific mechanisms by which regenerative grazing and pasture management achieve these results. Agriculturalists, consumers, and policy makers all have roles in making these benefits widespread, but in order to take effective action, they need a basic understanding of this science-based approach to raising cattle. Thus equipped, they can help make our regional food systems more secure and resilient, while protecting the ecoservices that are essential to maintaining a livable planet.

PART 1

Impacts of Regenerative Grazing

CHAPTER 1

Regional Resilience

We are going to have to return to the old questions about local nature, local carrying capacities, and local needs.

—WENDELL BERRY[1]

We have never been excited about football, but on February 26, 2020, Gillette Stadium became more to us than simply the home of the New England Patriots. That was the day we learned that this sixty-six thousand–capacity venue was going to buy our Big Picture Beef hamburger patties for their concession stands. For over a year we had made sales presentations and hosted meat-tasting sessions in the hope of opening more wholesale accounts for our grass-fed beef company. Now we had landed a big one.

More good news followed. The very next day we received an email from our distributor, PFG (Performance Food Group), saying, "Dartmouth College also loved the flavor of the patties." Brown University wanted our burgers, as well. By the time February rolled into March, Big Picture Beef was set to service these three new accounts, plus Cornell University and the University of Massachusetts Amherst. Between the stadium and the universities, our weekly output of beef had swelled by an additional five thousand pounds. We were rolling.

It wasn't just the sales that thrilled us. We had gone into winter working with about a hundred Northeast family farms who were raising 100% grass-fed cattle for our label. Now, with the stadium and the colleges, we could bring on fifty-two additional farms. Down the road we envisioned not only exponential growth for Big Picture Beef but also widespread regenerative grazing in the Northeast. After years of laying groundwork, building capacity, and negotiating for processing space, our mission-driven business was on track for success.

But everyone knows what happened next. In March 2020, the COVID-19 pandemic abruptly closed colleges and sports arenas, along with everything else. All our restaurant and food service business vanished. Old accounts and new.

Our distributor scrambled to replace the old restaurant and cafeteria accounts with expanded grocery store business for our beef, and succeeded to a remarkable degree. Some of our existing grocery accounts tripled, as people started preparing all their meals at home. But there was another reason for the big orders from grocery stores. The pandemic had revealed a startling fact about our nation's access to food: a handful of giant corporations control our country's meat supply.

In the United States, four corporations operate twenty-four huge slaughter and processing plants that control 85 percent of the finished beef market.[2] Their assembly-line employees work in close proximity; in early 2020, many contracted COVID-19 and some died.[3] In May 2020, *USA Today* reported, "In all, at least thirty-eight meatpacking plants have ceased operations at some point since the start of the coronavirus pandemic. All closed for at least a day. Some have stayed closed for weeks."[4]

Crisis Upticks

In some parts of the country, the closures left meat cases empty, and the beef shortage boosted online sales. The American Grassfed Association reported a big uptick in grass-fed beef sales. The association's executive director Carrie Balkcom was quoted online saying, "Some of our larger producers have seen online sales increase 500 to 1,200 percent."[5]

Big Picture Beef doesn't offer online sales, but here in the Northeast, the COVID-19 crisis seemed to boost public interest in local farms and healthy food in general. (This is not to imply that "local" and "healthy" always go together; the pandemic spurred interest in *both* local food and healthy food.)

This was the second time we have seen a public health crisis boost sales of grass-fed beef. The first time was in the early years of the new millennium. Ridge was the founding director of the New England Livestock Alliance, where he worked with farmers in the region to raise and market healthy meat. At that time, farmers producing grass-fed beef relied on sales from

THE LEARNING CURVE

When the Time Is Right

In 2003, even people who knew that grass-fed beef was healthier than grain-fed weren't clamoring for it; the palatability of the meat was variable.

But Roger Fortin at Little Alaska Farm in Wales, Maine, was producing terrific beef for a brand I had developed. So to prove that 100% grass-fed beef can be a gourmet food, I entered some of Roger's steaks in a *Wine Spectator* magazine beef contest. All the other entries were grain-fed. The competition was stiff, including such well-known brands as Lobel's and Niman Ranch. So we were thrilled when our entry for filet mignon won first place.[6]

But apparently the word about our big win didn't reach Northeast wholesale beef buyers. Selling grass-fed beef to stores and restaurants still felt like a push on a rope.

Then at the end of 2003, a national health crisis turned the marketing challenge around. I had no sooner read that the first case of mad cow disease had been reported in the United States when the phone rang. It was Boston TV Channel 7 wanting to know if grass-fed cattle could get mad cow. I said no, and explained that cows could get this disease only if they were fed parts of other cows. A camera crew came right out and filmed me for a news spot. The day after the clip aired on TV, the owner of a Boston deli rang me up—a guy I had approached about buying grass-fed beef with no success. As soon as I realized who was calling, I said, "I bet you've decided to buy my beef!" Yup. He was my first Boston customer.

—Ridge

farmers markets or their freezer trade (people who regularly buy from farmers to restock their home freezers); it was difficult to get wholesale buyers to sample grass-fed beef.

But that changed with the outbreak of disease: bovine spongiform encephalopathy (BSE), commonly known as "mad cow." Both people and cattle get this disease by consuming infected beef products. Mad cow damages the central nervous system. There is no cure; it is always fatal.

In December 2003, the first known case in this country received considerable media coverage.[7] Naturally the public was concerned. But as people began learning that cattle raised entirely on pasture plants cannot get mad cow, they started asking their grocers for 100% grass-fed beef. And they never stopped asking. The market for grass-fed beef has continued to grow.[8]

By the Numbers

Grass-fed beef has seen dramatic increases in sales over the last several years.

- Store sales in dollars of grass-fed beef have increased by 30 percent from 2019 to 2020, and pounds sold have increased 28 percent from 2019 to 2020.[9]
- The US grass-fed beef market increased from $17 million in retail sales in 2012 to $480 million in 2019, and this trend is expected to continue in the years ahead.[10]
- In 2015, the labeled grass-fed beef industry—including both imported and domestic grass-fed beef—was a $1 billion industry. In 2019, four years later, it had reached $1.6 billion.[11]
- Looking just at retail sales (and excluding restaurants), the grass-fed market rose by 16 percent for the year of research, ending February 24, 2019, and that followed an increase of 21 percent the year before that.[12]
- According to a report released on March 29, 2021, by Research and Markets, the North American grass-fed market is expected to hold the lion's share of the global grass-fed beef market (47.5 percent) by 2025.[13]

Market Outlook

One way to gauge the growth of customer demand for 100% grass-fed beef since the early years of the millennium—and prior to COVID-19—is to look at the response from businesses that sell beef.[14]

- Major meat processors have entered the grass-fed beef market sector (for example, JBS Foods, Tyson Foods, Cargill, National Beef Packing Company).
- Major food service distributors offer grass-fed beef (for example, Performance Food Group, US Foods, Aramark).
- Nearly all major retailers carry it (for example, Whole Foods, Sprouts Farmers Market, Kroger, Walmart, Target, Costco, Albertsons/Safeway).
- Mainstream restaurants have it on their menus (for example, Chipotle Mexican Grill, Carl's Jr./Hardee's, Outback Steakhouse, Chili's).
- Well-established, branded food companies have added grass-fed beef to their market offerings (for example, Meyer Natural Foods, Niman Ranch, Strauss Meats, Omaha Steaks, Allen Brothers, Maverick Ranch Meats, and Nolan Ryan Beef).
- Branded grass-fed programs (whose labeled packages are in grocery store meat cases) have grown their sales (for example, Big Picture Beef, Thousand Hills Cattle Company in Minnesota).
- Many of the e-commerce companies that have exploded into the marketplace feature grass-fed beef (for example, ButcherBox, Verde Farms, Omaha Steaks, DeBragga, Gourmet Beef).

Various reasons are offered for the growth in consumer interest: environmental benefits, animal welfare, and health issues (that is, not only benefits from eating grass-fed beef but also concerns about eating conventional beef). Midan Marketing, a Chicago-based strategic marketing firm, notes, "In just a few short years, the U.S. population has changed significantly, driven by younger generations with increasingly diverse ethnic profiles and attitudes. Topics like grass-fed beef, no antibiotics ever . . . are now at the forefront.[15]

Why Northeast Cattle Are Heading West

The default system of sending cattle west to be fattened in feedlots on corn has developed over years for various reasons that were once compelling.

- The association of beef with Western states began with the cattle drives of the mid-1800s from Texas to market towns in Kansas, including Abilene, Wichita, and Dodge City. These drives became unnecessary as more railroad tracks were laid in the West and Midwest in the last half of the nineteenth century and cattle could be transported by train. But these legendary events linger in the public imagination.
- The twentieth century brought additional, transformative developments that led to the feedlot system: hybridized corn varieties, synthetic fertilizers and biocides, and self-propelled combines for harvesting grain.[16] (The flat, wide-open spaces of the Great Plains are perfectly suited to giant combines.)
- Federal farm subsidies, which began in the 1930s and continue to the present day, make corn cheap feed for livestock. Therefore grain-fed cattle cost less to raise than grass-fed, and grain-fed beef costs less to buy. This is a huge economic challenge for grass-fed producers. But if those subsidies were eliminated so that the true cost of corn-fed beef were reflected in the price at checkout, the situation would be reversed; fattening cattle on pasture would be the norm because it would be the lower-cost mode of production. (See chapter 9 for more on subsidies.)
- By the 1950s the Great Plains was producing surplus corn, so ranchers began confining cattle and fattening them on this inexpensive grain. Cattle feedlots (CAFOs—confined animal feeding operations) are still concentrated in the Great Plains. They are also located in parts of the Corn Belt and the Southwest, with some feedlots in the Pacific Northwest.[17]

With history having linked "corn-fed" and "cattle" in the public mind, the mistaken assumption that corn-fed beef is more flavorful dies hard. This notion can be dispelled by a taste test (or a contest—see sidebar "When the Time Is Right," page 17).

Missed Opportunity

In light of burgeoning consumer interest in grass-fed beef, it is ironic that "the vast majority" of the 433,000 beef cattle born in the Northeast each year are purchased from farmers by cattle dealers and trucked out of the region to be fattened in Western feedlots on corn—not grass—as Mike Baker, an extension specialist at Cornell, told us.[18] After slaughter and processing, the beef from these Northeast-born cattle is sold all over the United States.

Although the superstores and supermarkets carry grass-fed beef, 75 to 80 percent of it comes from overseas, mostly from Australia, Uruguay, Argentina, and New Zealand. The country of origin is not apparent from the text on the packages, which says "Product of USA," only because the meat was minimally processed and packaged here.[19] This deceptive packaging has been legal since the repeal of a law that required beef packaging to include the country of origin (see chapter 9).

The other little-known fact is that very little imported beef is produced by best practices; that is, the regenerative grazing methods that restore degraded land and sequester carbon. As a result, while consumers are paying for environmental benefits that they associate with grass-fed beef, the beef in their shopping cart may not be produced in a way that delivers those benefits (see sidebar "No Graziers, No Paddocks: No Regenerative Grazing," page 43). Because the continuous grazing method used by many producers overseas requires very little labor, it is an inexpensive way to feed cattle; even with the cost of shipping meat to the United States, the imported product is usually less expensive than grass-fed beef produced regeneratively in the United States.

Grass-fed beef producers here feel that if people understood that the beef they are buying was *not* raised in the United States and that the production methods were *not* benefiting the environment, many of them would seek out a local, regional, or domestic brand despite a slightly higher price. With all the large grocery stores in the Northeast carrying grass-fed beef, there is clearly a market for it here.

Every truckload of cattle heading out of our region represents a loss for Northeast resilience. The opportunity costs that leave our region with the animals add up to a lot of lost value for the Northeast:

- the sale price for the cattle
- jobs for Northeast graziers who could fatten them on pasture
- ecosystem benefits from regenerative grazing
- more business for our packing plants
- marketable material from their slaughter (hide, bones, organs, glands, and so forth)
- the sale price of packaged grass-fed beef
- most important, healthy, delicious, protein: 100% grass-fed Northeast beef for Northeast consumers

Even though the Northeast has good soils, abundant rainfall, ample pastureland, and grass-fed beef producers wanting to expand their sales, unfortunately, the supply of grass-fed beef has not caught up with the demand. This is true in other parts of the country, as well. We will touch on the various difficulties in building the domestic supply, but let's start with the initial challenge we faced in starting a 100% grass-fed beef company here in the Northeast: meeting certain requirements of the wholesale marketplace. This was the nut we had to crack to open big accounts for beef produced in our region.

Meeting Wholesale Requirements

Wholesale vendors need both high volume and high quality consistently. These buyers will pay more for grass-fed than grain-fed beef *if* the price fits their budget, *if* they can get enough volume, *if* the meat is tender and tasty, and *if* the high quality is consistent—tender and tasty every time. For years,

Northeast farmers and farm cooperatives have bumped up against these requirements. Individual farms cannot provide the volume, and cooperatives have struggled with consistent quality because their member farms have a variety of conditions, livestock breeds, and management practices, and these variables are reflected in the meat.

But the steady growth of Big Picture Beef prior to the pandemic demonstrates that the Northeast *can* produce its own grass-fed beef. And other regions around the country can do so, as well. How? By giving local farmers/ ranchers the opportunity to make a profit raising grass-fed cattle. Finally, in the Northeast, this opportunity is at hand because we have developed a system that solves the high volume–consistent quality conundrum and enables farmers throughout the region to raise cattle for high-paying wholesale markets: stores, restaurants, and institutions. To build a regional supply of grass-fed cattle that will yield tender, tasty beef every time, the two distinct phases of production that we mentioned in the introduction must be addressed: (1) raising young stock and (2) fattening the cattle for market. In our program, for the first phase, cow-calf producers throughout the Northeast—operating independently—raise grass-fed young stock. For the second phase, we aggregate a number of these small herds to form a large herd (not unlike a herd of buffalo) that a skilled grazier fattens on a large finishing farm on grass and forage, without grain of any kind. The aggregation of the small herds provides volume, and the skilled finishing of the combined herds provides consistently great flavor.

When cattle producers adopt the regenerative practices necessary to meet the high standards of wholesale markets for grass-fed beef, they see improvements in their herds, an increase in soil organic matter (material that was once living), and lush pastures that stay green during dry seasons. Their neighbors also see these improvements.

For those farmers who don't have the time or the available labor to cultivate a freezer trade or take their meat to farmers markets, a big incentive to raise cattle for the wholesale grass-fed beef market is getting a fair price for their animals on a regular basis. Northeast producers who sell cattle to the commodity market are paid for pounds, not for quality; at an auction, 100% grass-fed beef sells for the same per-pound price as conventionally raised cattle. But the Northeast wholesale accounts that we have opened are willing to pay more for beef raised in our region on pasture alone without corn, added hormones, or antibiotics, so we have been able, in turn, to pay participating farmers a commensurately higher rate, amounting to 15 to 20 percent per pound over the commodity price.

As more farmers/ranchers learn that the price for 100% grass-fed beef is higher than the commodity price, many change the way they raise cattle in order to produce higher quality and benefit from the opportunity. This has already begun to happen in the Northeast. Every week we hear from additional farmers who want to participate.

In addition to getting paid for the quality of their cattle, the two-phase aggregation model offers farmers a number of other advantages:

- They can choose whether to raise young stock or to finish, depending on their interests, skills, and available pasture resources.
- They don't have to manage slaughter, processing, or marketing.
- They don't need to educate consumers on the benefits of grass-fed beef.
- And perhaps most important, they can access wholesale markets.

While the grass-fed beef movement is still in its infancy relative to all beef sales, this product and this production model offer a path to success for interested farmers and ranchers. According to the *Power of Meat 2019*, 54 percent of grocery shoppers would like to see more grass-fed beef in the stores.[20]

Net Profitability: Case Studies

The methods that we require of our partner farms here in the Northeast are applicable to or adaptable for other regions. Farmers and ranchers all around North America have made their grass-fed beef operations profitable. Typically, these producers did not turn to regenerative farming to combat climate change or for other environmental benefit; they adopted these practices to make a living doing what they enjoy, or to get out of debt, or to avoid financial ruin.

Richard Teague, a research ecologist from Texas A&M who has studied many farms and ranches around the United States, reports that successful grass-fed beef producers typically make the transition from conventional to regenerative methods by experimenting. For example, if they had been using a large quantity of chemicals, they don't eliminate them all in one year. The first year they might cut their applications in half, and then halve them again the second year. By the third year they might eliminate the chemicals—and that expense—entirely. Teague explains, "Profitability is not by buying stuff and putting it in the soil; it's by looking after the biology and relying on improved profits that accrue from the soil functioning better."[21]

Following are a few of many examples of farms and ranches that have made regenerative grazing profitable.

Dark Branch Farms, Kenansville, North Carolina

A few years ago, a dramatic success story unfolded at Dark Branch Farms in North Carolina's coastal plains, where Adam Grady and his family raise cattle, pigs, and field crops on their 1,200 acres of arable land.[22] The family has been farming there for many generations but only began their transition from conventional to regenerative methods in 2016. Just two years later in 2018, Hurricane Florence hit the area with thirty-five inches of torrential rain that covered the farm with eight-to-nine inches of floodwaters. The Gradys lost their soybean crop and all of their cover crops, and their perennial pastures were significantly damaged. But almost immediately, Adam Grady could see that those two years of fostering soil biology with regenerative practices had paid off. The soil structure had improved so dramatically that the farm's fields dried out quickly. Grady was able to plant them with cover crops just two weeks after the floodwaters receded, while his neighbor's fields were still too wet to support the weight of a tractor. Within a month the Dark Branch Farms' fields were green again, whereas neighboring fields were brown and dead.

Grady is now a devoted advocate of regenerative methods. Combining no-till with multispecies cover crops and then grazing those cover crops with livestock has continued to be a winning formula for him.

Brown's Ranch, Bismarck, North Dakota

A thousand miles away from the Grady farm is Brown's Ranch in North Dakota, owned and operated by Gabe Brown and his family.[23] Brown is the author of *Dirt to Soil*, his account of the transition from conventional to regenerative farming on his five thousand–acre ranch outside of Bismarck, North Dakota. His regenerative practices have increased the populations of soil microbes in his fields, with increases in the yields of his field crops year after year since the 1990s. He notes that bringing grazing animals into his fields has increased both the health of the soil and the profitability of the farm.

Brown began using the new methods when a series of natural disasters left him so overextended financially that he could not afford the chemical inputs that he had used previously. He says that his cost savings from eliminating chemical fertilizer, combined with increased yields from better-functioning soil, has made his ranch the most profitable farm in his county.

Triple T, North-Central South Dakota

Rancher Roy Thompson also reduced chemical inputs after switching from conventional methods to regenerative farming and raising grass-fed beef cattle on his South Dakota ranch.[24] Only a few years after changing his farming practices, he experienced the financial relief of eliminating his fertilizer cost entirely:

> I did not spend $4,500 on drop tubes [a fertilizer delivery system] and didn't put an ounce of fertilizer down. I just got my tissue samples back today and we are maxed on the high side for N [nitrogen]. What a savings! I would have spent another $10,000 just because that's what I always would have done. . . . This has helped me to be profitable in a year when it looked very unlikely.

Meeting Place Pastures, Cornwall, Vermont

Cheryl and Marc Cesario operate a 1,200-acre farm in Cornwall, Vermont.[25] They contract-graze both dairy and beef cattle on organic-certified acreage for a variety of customers and markets, and describe their operation as "trusted grazing services for livelihood, livestock and land." They also graze sheep and sell both beef and lamb directly to consumers. Cheryl is a Grazing Outreach Professional with the University of Vermont Extension in Middlebury.

Marc says, "Since 2009 we have converted 338 acres of corn land to perennial pasture; 310 acres of hay land to pasture; and 171 acres of set-stocked pasture [conventional grazing] to Adaptive Grazing."

The Cesarios adopted regenerative grazing because they believe it is a more economical way to raise cattle. Soil and water conservation practices have reduced their dependence on off-farm inputs, improved the health of their soil, protected water quality, and saved time and money.

————————

While we cite peer-reviewed, scientific studies to document the claims we make, successful operations such as these bring the published reports to life. One can debate endlessly about the methodology or conclusions of any given study, but the arguments, pro or con, are less compelling to us than lush, green pastures that have withstood droughts or floods, healthy cattle that are flourishing on grass and forage with no grain, and environmental practices that are profitable on operating farms.

From Marketing Niche to Mainstream?

We believe that 100% grass-fed beef can revive rural economies in the Northeast and elsewhere around North America. Why the optimism? Consumers all over the country are demanding grass-fed beef; it's a rapidly growing sector of the beef market.[26] While some customers want this beef because of its rich flavor, many are buying it for reasons that go beyond great taste. For the savvy twenty-first-century shopper, 100% grass-fed beef is an all-around winner.

- Grass-fed beef production offers big benefits for the environment, including a net climate benefit.[27] (See chapter 3.)
- Beef from cattle raised and fattened entirely on pasture—no grain ever—is a healthy source of protein for humans.[28] (See chapter 4.)
- If the beef is sold in the same region where it is produced, add "local" to this product's story. For years consumers have expressed a desire to reduce the "food miles" that their diet represents, and to keep their food dollars in the region where they live.

In less than two decades, we've seen the interest in this environmentally friendly beef grow rapidly. But for the grass-fed beef industry to continue on this growth trajectory, there are several challenges to be met.

CAPITAL. For the grass-fed beef industry to flourish and become mainstream, producers need capital for developing infrastructure—particularly for finishing. They also need working capital. The latter should be a banking function based on an understanding of beef production cash flow.

REGENERATIVE GRAZING METHODS FOR ALL GRASS-FED BEEF PRODUCTION. Raising grass-fed cattle calls for regenerative grazing and pasture management, which means that an experienced grazier moves the cattle through a rotation of paddocks (free of biocides and chemical fertilizers), taking care that each plot has sufficient time to regrow completely before being grazed again. We need many more graziers with the training and experience to know when a paddock has been rejuvenated and is ready for cattle.

A LARGE SUPPLY OF 100% GRASS-FED CATTLE OF CONSISTENT HIGH QUALITY. All grass-fed beef cattle need to be fully finished (fattened) on pasture with no grain. Intramuscular fat means flavor, and 100% grass-fed beef has a beneficial ratio of omega-6 fatty acids to omega-3 fatty acids that is important for health. Chapter 6 explains how to achieve high-quality beef consistently on a scale to supply wholesale accounts.

AGGREGATION OF GRASS-FED HERDS TO REACH ECONOMIES OF SCALE. Small-scale, grass-fed beef companies face economic realities that larger producers simply bypass. For example, the cost of trucking eight animals might run $225 per head, whereas the cost of trucking thirty-eight animals might be $89 per head. The same kind of math applies to the costs of slaughter and processing. Aggregation allows small- or medium-sized farms and meat businesses to access wholesale markets and thereby reach some economies of scale. Note that our model for developing regional programs involves aggregation prior to finishing, as well as raising young stock to a protocol.

WIDESPREAD USE OF TECHNOLOGY THAT PROVIDES TRANSPARENCY AND TRACEABILITY. Consumers want to be sure that the grass-fed beef in their cart is the real thing. From our perspective, "real" means that the cattle are grazed regeneratively their entire lives on pasture (with no chemical fertilizers, biocides, or GMOs), never ate grain, and received no growth hormones or antibiotics.

But any federally controlled standards for the marketing claim "grass-fed" would be open to influence—and subsequent weakening—by lobbyists over time. For example, the government has certified "organic" food for years, but now this certification allows crops grown hydroponically without soil, and dairy animals that live on feedlots rather than pastures. This prompted a new certifying body called Real Organic to create an add-on label for certified products so that shoppers can identify vegetables and fruits that were grown in soil, and milk and meat that came from pastured animals.[29]

The number of certifying companies for meat seems to be increasing with the current interest in verifying regenerative practices. Most companies charge the producer a fee to verify their claims and then communicate the farm's compliance with a label of some sort. To delineate all of these companies and their certifying criteria is beyond the scope of this book.

Because branded programs typically publish their management protocols, currently the best assurance that "grass-fed beef" is what it's claimed to be may be traceability. We are developing a tracking system that will enable consumers to trace the meat in their shopping cart back to the farm where the animal was raised. The process begins by tagging a new calf with an electronic ear tag. At the time the ear tag is put in, a sample of DNA is captured, so right from the start there is documentation of the calf's sire and dam. But that is just the start; a cow is tracked even if it crosses a road to a new paddock.

In the EU, such traceability is the law.[30] At the end of the animal's life, a "passport" is printed out. By law, it must travel with the animal to the packer. There the meat from that animal is correlated to the documented history of its life with a scannable barcode, which then travels into the marketplace with the packages of beef, so that a shopper can scan the barcode with a phone. Because this system is required in Europe, in a restaurant in France you can ask, "Where did this meat come from?" and the server can produce a passport with date of birth, farm location, date of harvest, and more. By contrast, in the United States, where traceability is not required by law, often the server provides the name of a distributor, with no reference to the farm or its locale.

HIGH-QUALITY, AFFORDABLE SLAUGHTER AND PROCESSING FACILITIES IN EVERY REGION. We began this chapter describing how the outbreak of COVID-19 abruptly eliminated all our restaurant and food service accounts. But we recovered much of that loss with new or expanded grocery store accounts and breathed easier—for a few moments. Soon we found that the loss of sales was only part of the pandemic impact on meat businesses; another issue was slaughter and processing. The problem became acute when shoppers faced meat shortages in the supermarkets and turned to local sources. That sounds like a good thing for a regional business, but in response to the demand for local beef, our Northeast farmers sent all the stock they had available to the small- and medium-sized packing plants—the only options in our region—and these plants were overwhelmed.

The shutdowns of the big plants that made headlines were only the most visible of the pandemic-related problems that strained local as well as national slaughter infrastructure. All packing plants, including the small ones, lost employees to COVID-19 or to the fear of COVID-19 or because school closings had upended their workers' lives. In some Northeast plants, slaughter slots backed up for two years. Although our business had been a regular customer of a midsized packing plant—which we had patronized from the beginning of our company's operation—our packer slashed the number of slaughter slots available to us in order to accommodate a bigger business that had invested in the plant.

We have since explored a number of solutions with variable success. The challenge is ongoing. (Chapter 8 describes processing problems that both reflect and impact larger issues in our society.)

––––––––––

The good news is that regenerative principles are already proven, and they are being applied successfully and profitably on working farms and ranches all around North America. The methodology for producing 100% grass-fed beef can be adapted to a variety of climatic conditions, from subzero winters to desert environments (see chapter 3). Our country has the capacity to shorten beef supply chains region by region so that everyone has access to nutrients from healthy meat, even in times of crisis.

CHAPTER 2

The Empty Breadbasket

*The value of industrially managed land depends on false
and brittle economies, such as access to government subsidies and the availability of cheap industrial fertilizer.*

—RICARDO SALVADOR[1]

T he Corn Belt takes up a big swath of our country: from Pennsylvania through Nebraska. The fifteen billion bushels of corn the
United States produces each year go mostly to animal feed, ethanol
and distillers' grain, high-fructose corn syrup, and other sweeteners.[2] But
more sobering than the issue of expending resources on this particular crop
is the scope of this monoculture. Planting the same annual crop year after
year causes a buildup and spread of pests and diseases, compaction, and soil
erosion from plowing, wind, and rain.

Adding to the assault on the land is Roundup, a biocide developed by
Monsanto in 1974 that farmers typically spray on their cornfields to eliminate weeds.[3] Seed corn is genetically modified to resist the herbicide, and is
advertised as Roundup Ready.

The active ingredient in Roundup is glyphosate, which is powerful stuff.
It kills by stopping a specific enzyme pathway that most plants and microorganisms need for survival. Roundup is banned or restricted in a dozen
countries in Europe and the Middle East. As it is present virtually everywhere
in the food chain, it is regularly ingested.[4] The World Health Organization
has classified glyphosate as a probable human carcinogen.

Regarding the impact of glyphosate on farmland, both peer-reviewed
studies and reports from the field confirm that this herbicide is harmful to
soil life, including mycorrhizal fungi, beneficial microbes that live in, on, and

around plant roots.[5] Soil is a mixture of rock-derived minerals, organic matter, water, gas, and countless microbes; just a teaspoon of soil includes more living organisms than there are people on Earth.[6] The presence or absence of these organisms is what makes soil healthy or not. Soil life, though largely unseen, is at the heart of the movement to raise and fatten cattle on grass and forage instead of corn.

THE LEARNING CURVE

Caution: Monoculture Ahead

Driving west through Illinois one spring, I was struck anew that US agriculture has gone dangerously awry. On either side of the highway for mile after mile were bare fields soon to be planted with corn.

We all marvel at the way plants manage to grow almost anywhere, even through cracks in concrete, right? Along that Illinois highway, I saw firsthand what was coming up in those unplanted fields. Nothing! I could not see a single weed; the soil appeared to be devoid of life. In contrast, the highway's median strip was teeming with grass and other vegetation. But not in those Illinois cornfields, not in the "breadbasket of the world."

The cornfields had been sprayed with Monsanto's Roundup to eliminate weeds that would compete with the corn.

In contrast, farmers working in harmony with natural soil systems have noted that weeds are less of a problem on acreage with a robust population of fungi. A study evaluating fungi for weed control concluded that fungi's suppressive effect on the growth of some weeds, "might be of particular interest to more sustainable farming systems, where weed management to tolerable levels, rather than total weed eradication, is the prevailing strategy."[7]

—Ridge

In recent decades, soil scientists, ecologists, and agriculturalists in the United States and around the world have come to understand that the interdependent roles of soil microbes and grazing ruminants can address some of our most confounding global problems. Now we know the specific mechanisms by which well-managed grazing not only fattens cattle but also fosters populations of beneficial microbes. Soil microbes are key players in natural processes that can restore degraded farmland, protect against droughts and floods, increase biodiversity, and combat climate change through carbon sequestration.[8]

But these microbes are threatened by destructive farming practices, as noted in the *Atlantic*:

> We are now at a point where microbes that thrive in healthy soil have been largely rendered inactive or eliminated in most commercial agricultural lands; they are unable to do what they have done for hundreds of millions of years, to access, conserve, and cycle nutrients and water for plants and regulate the climate. . . . We need these tiny partners to help build a sustainable agricultural system, to stabilize our climate in an era of increasing drought and severe weather, and to maintain our very health and well-being.[9]

Unfortunately, planting monocultures and spraying glyphosate are not the only destructive activities associated with producing corn. Additional practices that degrade farmland include clearing land of vegetation, applying chemical fertilizers, and plowing.[10] Wait—*plowing*? Yes, some of these practices are so familiar that they seem benign. But now, after decades of erosion and decreasing fertility, soil scientists have documented the ways in which many conventional farming practices, such as plowing, actually work against natural processes that are essential for healthy soil and for all the species that depend upon it, including humans. More about that to come.

Another compelling reason to avoid feeding corn to cattle bridges the topics of ecology and bovine health. Corn is not a natural food for ruminant animals. All ruminants have four-part digestive systems. The fourth part is called the rumen, and it allows the animal to digest cellulose. Because of this digestive system, the natural diet for cattle is grass or hay (which is dried grass), and other pasture plants. Over time and in quantity, corn sickens cattle, which is why feedlots dose them with antibiotics preventively to keep them alive and gaining weight until slaughter.[11] When cattle are allowed to feed themselves in a pasture—where they can also rest, socialize, and care for their young according to their instincts as bovines—they tend to be healthy naturally, and the pasture flourishes, as well.

Despite the well-known problems associated with growing corn to feed cattle—and notwithstanding the fledgling grass-fed beef movement—the corn-based, feedlot production of cattle continues full tilt. As we have noted, most of the cattle born here in the Northeast are trucked to Western feed-lots for corn fattening, with all the diesel emissions and waste of fuel that transporting hundreds of thousands of cattle entails.[12] At the feedlots, the environmental degradation includes the runoff of urine and manure, which causes nutrient loading, and the climate-changing methane emissions from both the grain-fed cattle and the accumulation of manure in lagoons.

The environmental destruction from corn production is also associated with other crops that are replanted each year, such as soybeans and peas. (Remember this when you hear someone praising meatless burgers that include these industrially produced vegetables.) But we're focusing on corn production here because (1) it uses more land area of the United States than any other crop (land that could be utilized for regenerative grazing); (2) nearly half of the corn is grown to feed livestock; and (3) corn represents 60 to 85 percent of feedlot cattle's diet.[13]

The Big Picture

Consider that a commercial cornfield is an ecosystem, the technical term being *agroecosystem*. Any ecosystem includes air, water, soil, plants, animals, and microbes such as fungi and bacteria (which are alive but are neither plants nor animals). All these things, both living and nonliving, are interdependent. As we shall see, even microbes communicate with chemical signals and carry out complex biochemical processes together. Therefore, everything farmers do in a cornfield to plant, fertilize, cultivate, and harvest corn impacts the whole ecosystem one way or another.

Though this can get complicated, farmers, concerned consumers, and policy makers don't need to be biochemists to understand how the harm done by our nation's corn production ripples far from the fields where the corn is grown.

Plowing, for example, contributes to the seemingly unrelated issue of carbon dioxide building up in the atmosphere. For perspective, consider that for many years before the beginnings of agriculture, the carbon cycle was humming along, with the rate of carbon drawdown into the soil exceeding the rate of carbon release back to the atmosphere. Then about ten thousand years ago, humans began to reverse this. Carbon is released to the atmosphere through oxidation of soil carbon, a process akin to rusting. In the last seventy years, coinciding with the advent of industrial agriculture, we have accelerated the loss of carbon from our soils by removing vegetation covering the land and plowing, thus exposing soil carbon to the air, where it is oxidized and becomes carbon dioxide.

Soil *tilth*, a crumbly condition, suggests that soil is suitable for planting and may indicate that the soil is well aggregated, meaning that it has a carbon-based "sponge" that can hold water. For generations, agriculturalists thought that tilth was created by plowing, but now we understand how plowing undermines soil fertility and structure. Scientists estimate that most agricultural soils have lost 30 to 75 percent of their soil organic carbon since tillage-based farming began. Industrial agriculture accelerated these losses in the twentieth century, largely since World War II.[14]

Recently, farmers have begun to eliminate or decrease tillage on their cropland (a recent survey indicates that now approximately a third of acreage planted is left untilled), but most of these farmers use glyphosate to kill their cover crops before sowing seeds with no-till planters.[15]

While plowing is a very old practice, synthetic fertilizers, pesticides, and herbicides have been used routinely since the end of World War II, when chemical companies teamed up with agricultural experts to use the chemicals of warfare on farmland to increase fertility and eliminate weed and insect pests.[16] Chemical fertilizer production depends on continued fossil fuel extraction.[17]

Ironically, high-input agriculture in some areas has led to a *decrease* in yields because of the harm of tillage, chemical fertilizers, and biocides to soil structure and soil organisms.[18]

Is Plant-Based Better?

A calculation of the environmental impact from conventional beef production must include all the harmful impacts of raising the corn for their feed (chemical fertilizers and biocides, monoculture plantings, oxidation of soil carbon, and the fossil fuels that power the machines). However, when all this damage is tallied, many people lump 100% grass-fed beef together with corn-fed beef, even though grass-fed cattle don't eat a single kernel of corn. Often commentators point to the toll of conventional beef production and simply urge people to eat less meat; their articles, books, and blogs either don't mention grass-fed beef or else perpetuate misinformation about its production.[19]

Although people choose a vegetarian or vegan diet for many reasons, in terms of the environmental impact of food production, industrial vegetable production is not a viable alternative to industrial meat production. The popular misconception that plant-based products are always better for the environment than animal products has been disproven by peer-reviewed studies, both those measuring the harm caused by conventional crop farming, and those measuring the benefits of regenerative grazing to address that harm.[20] An abundance of evidence links conventional vegetable farming to the destruction of beneficial microbes, a decline in nutrient cycling, the loss of soil carbon, flooding, erosion, pollution of waterways, the death of wildlife, and high greenhouse gas emissions.[21]

Fortunately, there are proven, alternative ways to produce beef that work with beneficial systems in the natural world. We will describe these practices in the next chapter, but first let's be specific about which agricultural practices work *against* nature's systems and therefore should be replaced.

Rites of Destruction

The following list describes destructive practices and outcomes that occur every spring in every region of the United States where corn or other annual crops are grown:

- Crops are planted in vast monocultures. In nature, one species never dominates to the total exclusion of others. Single-species planting creates the buildup of pests and diseases.
- When fields are plowed and planted, the soil is bare. You seldom see bare soil in nature because photosynthesis cannot take place in bare soils. We need as many energy-capturing, carbon-pumping plants as possible.

- Leaving soil bare also causes erosion by exposing the surface to wind, rain, and flooding. The average rate of soil loss in farmland is 5.8 tons per acre per year.[22]
- Plowing increases the volume of carbon that is oxidized and therefore lost to our soils and released to the atmosphere. Soil carbon becomes carbon dioxide and contributes to climate change.
- Plowing, or similar soil disturbance, also upends soil structure, compacts the soil, and compromises microbes that are needed for healthy soil function.
- Application of herbicides and pesticides compromises or destroys soil microbes, thereby inhibiting beneficial biological activity in the soil.
- Chemical fertilizers—nitrogen or water-soluble phosphorus—on farm fields interfere with the natural process whereby plants give microbes carbon in "exchange" for nutrients, a critical process that we describe in the next chapter.
- In waterlogged soil, nitrogen fertilizer produces nitrous oxide, an extremely potent greenhouse gas, which rises into the atmosphere. Nitrous oxide is approximately three hundred times more heat-trapping than carbon dioxide.[23]
- Also, excess fertilizer runs off fields causing pollution of groundwater and waterways. (See sidebar "The Fourth Global Crisis," page 38.)

Currently, 97 percent of the beef cattle produced in the United States are fed corn, which is grown in chemical-intensive monocultures where all of the practices described above are used every year.[24] The states bordering the upper Mississippi and Missouri Rivers that suffered billions of dollars' worth of damage from flooding in 2019 are also the states with the largest production of monocultures of corn and soy, much of which goes to feed livestock.[25]

By continuing to use destructive methods for growing corn and other annual crops, farmers work against natural processes that have evolved over millions of years. This erosion of soil systems has dire implications for all of us. A 2021 project from the University of Massachusetts used updated technology to estimate that the most fertile topsoil is gone from a third of all the land devoted to growing crops across the Upper Midwest.[30]

The Fourth Global Crisis

If you were asked to name our most critical global crisis, the first would likely be climate change, followed by deforestation or loss of biodiversity. But there is fourth global crisis. We have exceeded a fourth planetary "boundary" that hasn't received as much attention: nitrogen pollution.[26]

We all know about the dead zone in the Gulf of Mexico, where excess nitrogen stimulates so much aquatic plant growth that the resulting rotting mass suffocates other marine life. This excess nitrogen, pouring down the Mississippi River from Midwestern farm fields, is part of the grim saga of corn production. That this dead zone at its largest has extended for 8,800 square miles is bad enough, but the really scary news is that there are more than four hundred dead zones in the world's oceans.[27]

And nitrogen pollution is not limited to oceans; it also affects bodies of fresh water, including drinking water supplies. Whereas in the past phosphorous fertilizer was the targeted polluter, now cleanup crews routinely face toxic blue-green algae fueled by nitrogen pollution.

Nitrogen is getting deposited everywhere, including on land, where it is reducing plant diversity because the native species that are adapted to nutrient-poor soils are being displaced. A study last year in the *Proceedings of the National Academy of Sciences* examined more than fifteen thousand forest, woodland, grassland, and

Currently, forage, grasslands, and grazing lands constitute more than two-thirds of all agricultural land in the United States.[31] Avoiding conversion of this land to plowed cropland could help prevent increases in greenhouse gas emissions and other harmful impacts to the local and global environment.[32]

shrubland sites across the United States and found that a quarter of them have already exceeded the nitrogen levels associated with species loss.[28]

Some scientists say that if we don't halve the amount of nitrogen that we dump into the environment by midcentury, we will face epidemics of toxic tides, lifeless rivers, and dead oceans.

But the microbes fostered by regenerative grazing (as we will explain in chapter 3) offer at least a partial solution to this dire situation: They transform organic nitrogen in the soil (not usable by plants) into the mineral nitrogen that plants do need.

This capacity for soil microbes to supply mineral nitrogen to plants is hardly a secret, but nevertheless, given the amount of nitrogen fertilizer used on corn for livestock, researchers from the University of Victoria have offered this mind-boggling solution: laboratory cultured meat.[29] Their reasoning is that much less fertilizer would be needed for corn production if we reduced the global herd of cattle from 1.5 billion to the 30,000 cattle that would be used as the stem-cell donors needed to create artificial meat. (For more on the lab meat movement, see chapter 9.)

Fortunately, there is a simpler way to reduce nitrogen fertilizer use for corn crops that can solve many other problems at the same time: stop feeding corn to cattle and raise them entirely on pasture instead.

And if practitioners and policy makers make full use of discoveries made by soil scientists and ecologists in recent decades, which we will detail in the next chapter, they can correct our disastrous course.

CHAPTER 3

Cattle as Global Heroes

Shifting from feedlot farming to rotational grazing is one of the few changes we could make that's on the same scale as the problem of global warming. It won't do away with the need for radically cutting emissions, but it could help get the car exhaust you emitted back in high school out of the atmosphere.

—BILL MCKIBBEN[1]

We've made broad claims about the benefits of regenerative grazing and pasture management. Now let's dig into those claims. Exactly how can grazing cattle—however well managed—rejuvenate degraded farmland? And how can grazing management increase the amount of carbon stored in our soils?

In this chapter we will explain how a set of regenerative practices can do both. We've already indicated that soil microorganisms are critical to regenerative agriculture. Without getting into biochemistry, we will outline the specific ways that microbes can not only help feed the world but can also help forestall the worst impacts of climate change. The two goals are closely linked, and we can reach them sooner and less expensively than you might think by fostering synergies among microbes, pasture plants, and cattle.

Nature has evolved intricate systems to provide nourishment for both plants and animals. Whereas humans have spent the last ten thousand years developing agricultural practices, plants make their own food through the process of photosynthesis. But there is more to plant nutrition than the process that we learned about in school. Photosynthesis is directly linked to a network of natural processes that collectively are responsible for nutrient cycling, which is the movement of nutrients from the physical environment to living organisms and back to the environment.

Since that sounds a bit abstract, let's start with something tangible and familiar: grass.

Feedback

To get the nutrition they need when they need it, grasses and other pasture plants use feedback mechanisms that are not well known beyond the scientific community but have important implications for humans regardless of their diet.

Here's how a feedback mechanism works in a pasture early in the grazing season. After a cow takes a bite of grass, that partially defoliated plant needs additional nutrients to regrow. In response to that need, the grass sends a chemical signal down to its roots to release some of the carbon compounds it has stored there.[2] (How this helps the plant regrow may not be apparent yet, but bear with us.) When these carbon compounds (sugary substances called *exudates*) are released by the roots, they attract soil microbes, notably fungi and bacteria.

What follows is a burst of microbial activity. In the simplest terms, various soil microbes do all the following:

- feast on carbon compounds (and each other)
- reproduce
- store some of the carbon from the roots as soil carbon
- bind soil minerals and organic matter together, providing soil structure
- supply mineral nitrogen and phosphorus to the plant for regrowth
- supply the plant with water and other soil nutrients that support growth[3]

Only in recent decades have scientists understood that the synergy of cattle, grassland plants, and soil microbes is the basis for creating and maintaining the health of a pasture, and in turn, the health of the animals that eat the pasture plants and the humans that eat the animals. It is also the basis for increasing the storage of carbon underground.

While the chemical processes involved are complex (and can also involve protozoa, nematodes, insects, and arachnids), the lead role in cycling soil nutrients belongs to mycorrhizal fungi, which live in, on, and around plant roots. These microbes "trade" the nutrients and water needed by the plant for the carbon in the plant's roots. This trade—nutrients for carbon—takes place as a two-way flow through the fungi's abundant hyphae, which are long filaments. These hyphae can reach beyond the spread of the plant's roots to sustenance that would otherwise be inaccessible to the plant.

Remember, all this activity is initiated by a cow cropping grass, which stimulates the grass roots to release carbon that attracts soil microbes; this in turn sets up a mutually beneficial exchange of carbon for nutrients that the plant needs. This is why grazing has been called a "jump start" for nutrient cycling.[4]

We must safeguard populations of fungi, bacteria, and other soil microbes, because a decline in the numbers of microbes around plant roots impedes the two-way exchange of nutrients for carbon. Fewer microbes means (1) a decrease in nutrients available for plants and (2) a decrease in carbon stored in soil—two good reasons to care for soil organisms rather than compromise them by the misguided agricultural practices described in the previous chapter.

When farmers and ranchers shift from those destructive practices to regenerative methods, they strengthen populations of microbes and other organisms in the soil food web. This approach works in harmony with nature's own systems, notably photosynthesis and the carbon-nutrient exchange that we described previously.

The Multiplier Effect

Some regenerative grazing practices, such as keeping the ground covered with growing plants, are essentially permaculture principles applied on a large scale, with cattle as an integral part of the system. As sustainable agricultural specialist Lee Rinehart notes:

> Perennial pastures, because of the lack of soil disturbance and their permanent cover, are higher in carbon and organic matter than tilled crop fields. This biological system has a stable habitat to conduct business, and the nutrient cycles can sustain themselves. However, we know that by adding livestock to the mix, we get a multiplier effect.[5]

THE LEARNING CURVE

No Graziers, No Paddocks: No Regenerative Grazing

My trip started in the middle of the night. A limousine made its way down our dirt road, picked me up, and took me to an airport outside of Boston, where I boarded a private jet at dawn. The owner of the jet had hired me as a consultant to evaluate the cattle and grazing operation at a large ranch he owned in Uruguay. We flew down together.

Uruguay's grasslands have the benefit of a rainy subtropical climate. My client's ranch was thousands of acres of deep, vegetative grass with a good perimeter fence. Although regenerative grazing is now practiced in some places in Uruguay, his cattle were managed conventionally, by *continuous grazing*, which means that the animals are simply turned loose in a large tract until they reach a marketable weight.

In Uruguay, with boundless grasslands, continuous grazing works well from the perspective of a ranch manager. Plenty of rain year-round ensures green grass, and ponds dotting the rangeland supply drinking water. The cattle thrive on their own. The verdant grassland in Uruguay was a stark contrast to the degraded rangeland I'd seen in the arid American West.

In most of the United States, where continuous grazing is also standard management, this method results in deterioration of the grassland. (Perhaps this is why "grazing" seems to be synonymous with "overgrazing" in the minds of many in this country.)

Conventional grazing means that the pasture (large or small) is not subdivided into paddocks. Typically, there is enough grass to eat and water to drink, so cattle fend for themselves with little or no labor or expense involved. But with no graziers and no paddocks, there is also no regenerative grazing. Without temporary paddocks within the large area, the cattle head for patches where their first choice of plants are growing. And for an entire season they keep returning to those patches to eat the first regrowth of their favorites, while other plants are left to spread and sometimes become invasive. Meanwhile, the favored patches become overgrazed, soil becomes compacted, and water cannot infiltrate. The resulting runoff of rain in turn leads to soil erosion and nutrient loss. Manure is distributed unevenly. Animals within the herd may not be receiving equal nourishment.

With vast tracks of land and a climate like Uruguay's, cattle can be well finished by this conventional approach, nevertheless. Unlike rangeland in the American West or Australia, which is very dry, the pastures in Uruguay are so large and so lush that the cattle can always move on to another green area. As I reported to my client, his cattle were in great shape.

Most of the imported grass-fed beef in US supermarkets today is produced by the continuous grazing method. The biggest markets for Uruguay's grass-fed beef are China and the United States.[6] Uruguay's beef sells in the United States for a very competitive price.

But while these cattle are pasture-raised, we think that the term *grass-fed beef* should be reserved for beef produced by adaptive multi-paddock grazing. Unlike continuous grazing, the regenerative approach fosters soil life, sequesters carbon, builds topsoil, and benefits the environment in multiple ways.

—*Ridge*

Ongoing grazing and pasture management is essential in order for beef producers to get the full benefit of raising cattle. Whereas the Great Plains had predators that kept the grazing buffalo herds moving across the prairie, in the regenerative model, the grazier (who could be a member of the farm family, the farm manager, or another experienced person) moves cattle through a rotation of paddocks, taking care that a grazed paddock fully regrows before the cattle return. This regrowth, stimulated by the previous grazing period, is not "rest"; during the period when the bovines are grazing elsewhere, the microbes and other organisms in the soil—though unseen—are multiplying and very actively participating in their respective tasks. Nutrients are transported, carbon is stored, roots grow long, and pasture plants grow tall again. The grazier moves the cattle at a pace that allows them to eat only the tops of the plants so that each paddock can successfully regrow. That means short grazing periods followed by longer regeneration periods—much longer in a dry climate. The grazier also adjusts animal numbers to match available forage. This method ensures that the cattle get optimal nutrition and the pasture flourishes both above and below ground.

This science-based approach is not just theoretical; it has been practiced for four decades in many parts of the world, with documented ecological and economic success in North America, Central America, South America, Hawaii, Australia, New Zealand, and southern Africa.[7]

Environmental Benefits

Through advances in soil science and ecology in recent decades, we have gained understanding of the mechanisms by which regenerative grazing can achieve all of the following:

- increased soil fertility
- stable hydrology and protection against erosion, floods, droughts, and desertification
- increased biodiversity above and below the soil line
- reduced greenhouse gas emissions: carbon dioxide, methane, and nitrous oxide
- more stored carbon

This list of environmental benefits may sound too good to be true, so let's examine the specific processes that bring about each of the five outcomes we've listed.

Increased Fertility

Fertility is a complex dynamic that is not achieved by simply adding fertilizer. Productive soil depends upon robust populations of living organisms, most of them microscopic, carrying out myriad functions that support plant growth and health. Cattle contribute more than manure to this dynamic—as we have explained, the grazing itself, properly managed, stimulates nutrient cycling. The following components combine to increase fertility:

EXUDATES. Soil ecologist Christine Jones points out that "every plant exudes its own blend of sugars, enzymes, phenols, amino acids, nucleic acids, auxins, gibberellins, and other biological compounds," but she often refers to exudates as "liquid carbon."[8] In the beginning of this chapter, we described the process by which grazing stimulates pasture plants to exude some of their stored carbon into the soil. These root exudates fertilize the soil either directly or via the bodies of microbes that have eaten the exudates.

MODERATE GRAZING. Graziers move the cattle from a paddock after they have eaten only the tops of the plants (ideally less than half). A short grazing period ensures that enough of the plant remains to conduct photosynthesis, and thus to regrow quickly and completely. In each paddock, *short* grazing periods and *long* recovery periods allow microbes to multiply, to make water and soil nutrients, including nitrogen and phosphorus, available to the pasture plants, and to stabilize carbon in humus. The regrown plants will be eaten by the cattle on their next rotation through the pasture, the timing of which is managed by the grazier, based on an assessment of conditions.

TRAMPLING VIA MOB GRAZING. Much of the vegetation that remains after grazing will be trampled, which is another benefit of the regenerative approach. This allows more sunlight to reach the photosynthesizing leaves. The cattle's hooves crush plant residues and press the organic material into the soil so that microbes have access to it quickly and decomposition can begin. Their hooves also bury seeds, and the hoofprints create mini-reservoirs that promote seed germination. Trampling is most effective if the cattle are numerous and moving close to one another. In evolution, this instinct to bunch was a defense against predators. Today, graziers achieve effective trampling by using a technique called mob grazing, which works with the cattle's herding instinct. Graziers keep the numbers of cattle high in proportion to the size of the paddock, and move them frequently to the next paddock so they always have fresh grass in front of them to eat. Careful monitoring and adaptation are important to achieve the benefits of trampling while avoiding undesired depletion of vegetation.

MANURE. While eating and trampling, cattle also deposit their manure and urine throughout the paddock, thus enriching the soil without the need for a manure spreader powered by fossil fuel. Dung beetles have a unique and important role in manure handling. Found on every continent except Antarctica, these insects quickly remove animal manure from the surface and bury it, subsequently using some of it for food and for incubating their eggs. Thus, the manure supplies the soil with both nutrients and beneficial microbes from the ruminant's digestive system. In addition, the dung beetles' tunnels help with rainfall infiltration.

DECOMPOSED SOIL LIFE. Dead roots, weed residues, green manures that decompose, and the remains of dead mycorrhizal fungi, bacteria, and other soil life are additional sources of fertility that accumulate in great quantity with well-managed grazing.

Land that has been grazed by regenerative practices can yield more than high-quality beef. The improved acreage can be used for crops as well as pasture for livestock, or it can be planted with cover crops that extend the grazing season. Crop yields will be more abundant than on land that has not been grazed regeneratively.[9] Whether the land is managed for crops or perennial pasture or a system that integrates livestock and cropping, it will yield significantly more biomass (plant material) than it did before it was improved by regenerative grazing. A research study led by Steven I. Apfelbaum compared conventional and regenerative grazing management and found that biomass was 300 percent higher with regenerative management.[10]

Scientists tell us that soil anywhere in the world is potentially fertile in terms of mineral nutrients.[11] The key is having a diversity of soil microbes—particularly fungi—to make the nutrients available to plants. Fertility is achieved by improving ecosystem function rather than by purchasing inputs to add to the soil. Richard Teague offers this time frame for seeing farmland rejuvenate:

Most tillage and chemical-based farming operations have diminished or destroyed the soil biota, but with management changes based on regenerative principles they can quickly recover. When you get rid of the elements suppressing key soil microbes and you start having a biodiverse mix of plants, you can get a response within a year or two in terms of soil regenerating.[12]

Stable Hydrology and Protection against Erosion, Droughts, and Floods

Droughts and floods are two sides of the same coin. Either can occur if soil lacks the capacity to hold water. Teague notes that regenerative grazing improves this capacity:

> Where regenerative AMP grazing has been practiced in semi-arid and arid lands for some time, plant productivity and biodiversity have been elevated, plant and litter cover have increased over the landscape, and nitrogen-fixing native leguminous plant species and pollinators have increased. This has resulted in re-perennialization of ephemeral streams and watershed function.[13]

In 1996, USDA soil scientist Sara F. Wright made an important discovery that has helped scientists understand how healthy soil structure allows rainwater to infiltrate soil and be retained or gradually released as needed.

Wright discovered the natural mechanism that gives soil a healthy "crumb." (Some people have likened healthy soil to chocolate cake.) She identified a sticky material in soil, which she named *glomalin*. This organic (carbon-based) material is produced by mycorrhizal fungi and is found on their long hyphae. In addition to sealing the hyphae (which allows them to transport liquid as part of the carbon-nutrient exchange), glomalin creates soil tilth by binding soil minerals—silt, sand, or clay—and organic matter together to form soil aggregates, which range from granules to pea-sized lumps. Well-structured soil is well aggregated, which means that the soil is better able to withstand erosion because the particles can't be torn apart by wind and rain.

While it is the organic matter, with its "glues and gums," that facilitates the formation of aggregates, their effectiveness in holding water is in large part due to the spaces between them.[14] Well-aggregated soil is said to be a *carbon sponge*. In an interview with *Acres U.S.A.* magazine, microbiologist Walter Jehne compared the carbon sponge to a cathedral:

> What's awe-inspiring about a cathedral are the voids and the ethereal spaces—the nothingness they create—not the bricks and the cement. Well-aggregated soil is like a cathedral. . . . About 66 percent of a healthy soil is just space, air—nothing—and that creates massive capacity for the sponge to hold water. . . . Creating these cathedrals—these spaces and surfaces—is fundamental for both soil hydrology and biofertility.[15]

Without the spaces that characterize good soil structure, rainwater accumulates on the surface, causing flooding, or it runs off, causing erosion. Even in the Northeast, where we generally have adequate rainfall on a yearly basis, a carbon sponge in the soil is necessary to keep grass green and growing during dry spells, and to buffer heavy rain events.

Jehne goes on to explain that the organic matter that improves the structure of the soil also helps cool the climate:

> Water vapor is uniquely powerful at absorbing heat. . . . When water evaporates from the land's surface or is transpired by vegetation or forests, that heat gets transferred from the Earth's surface up into the atmosphere, cooling the [Earth's] surface. . . . Most of that heat gets dissipated back out to space from the upper atmosphere. That process accounts for about 24 percent of the Earth's natural hydrological cooling. . . . We need green plants and organic matter in the soil to keep that whole hydrology working.[16]

Increased Biodiversity

One of the management goals of regenerative grazing is to increase biodiversity. Where grasslands are not grazed, light quickly favors the tallest plants, which results in a few tall species dominating. But grazing removes light as a limiting factor and enhances biodiversity as other plants can compete.[17] There is a positive correlation between plant diversity and microbial richness.[18] Microbial diversity is also enhanced by drilling a mixture of cover crops into the same pasture, which should include short, medium, and tall

From Desert to Diversity

When Alejandro Carrillo took over the operation of Las Damas Ranch, his father's cattle ranch in the Chihuahuan Desert, there was lots of bare ground and very little grass. This area in northern Mexico gets only eight to ten inches of rain a year. (The average annual rainfall here in Massachusetts is forty-eight inches.) Even with twenty-five thousand acres of rangeland, Carrillo could run only 250 cows, plus an additional 150 on leased land. And from March through July, he had to supplement grazing with corn-meal, cotton meal, and salt.[19]

Today this same struggling dryland ranch is verdant and profitable. The transformation began in 2002, when Carrillo decided to take a Holistic Management class in Chihuahua City. This course led to many more, and eventually Carrillo turned to regenerative grazing consultants, who helped him establish the grazing and pasture management practices he uses today. According to grazing expert Allen Williams, two keys to success at Las Damas have been (1) using temporary polywire fencing to form numerous small paddocks, which allow Carrillo to move his cattle frequently, and (2) developing the most productive areas first and then moving into the more

plants; both warm-season and cool-season plants; and both broad-leaved plants and grasses. Short grazing periods and long recovery periods will lead to myriad species of microbes performing the various functions of healthy soil systems as plant biodiversity increases. Rancher Gabe Brown has noted:

If you want a healthy, functioning ecosystem on your farm or ranch, you must provide a home and habitat for not only farm animals but also pollinators, predator insects, earthworms, and all of the microbiology that drive ecosystem function.[23]

degraded areas—desert—as the new practices began to change the land.[20]

"It took only one year to become profitable because we immediately stopped supplemental feeding, and combined herds and started moving the cattle more frequently. It took about three years to get enough forage that we could bring the cattle from the area we were leasing. That made the business even more profitable," says Carrillo. "Now we have multiple [species of] grasses, some taller than myself."[21]

The diversity of grasses has attracted many new species of birds. Andrew Rothman, director of the Migratory Bird Program at the American Bird Conservancy, has visited Las Damas Ranch several times. "It was pretty evident right away that the amount of grasses there was superior to pretty much any land we'd seen in the Chihuahuan Desert," Rothman says. "That land was different, with plenty of vegetation and forage for cattle. What's good forage for cattle is good habitat for birds."[22] Conservation groups are now working with Carrillo and neighboring ranchers to create a biological corridor that will offer protection for desert grassland birds so that their populations can rebound.

A diverse pasture also increases habitat for larger wildlife—invertebrates, mammals, birds—even while cattle are grazing.[24] (Regarding birds, see sidebar "From Desert to Diversity.") Diversity not only allows for multiple synergies among plants and animals, but is also a hedge against species extinction.

With billions of species on the planet and most of them microscopic, their relationships to one another are so complex that it is not possible to predict the impact of extinctions on the natural systems that sustain us. According to a United Nations report released in May 2019, plant and animal extinctions are occurring at a rate that is at least a thousand times faster now than in the time before humans were on Earth. This is alarming because the Earth's eco-systems include countless interdependent processes, such as those described in this 2019 opinion piece by Ferris Jabr:

> Trees, algae and other photosynthetic organisms produce most of the world's breathable oxygen, helping maintain it at a level high enough to support complex life, but not so high that Earth would erupt in flames at the slightest spark. Ocean plankton drive chemical cycles on which all other life depends and emit gases that increase cloud cover, altering global climate. Seaweed, coral reefs and shellfish store huge amounts of carbon, balance the ocean's chemistry and defend shore-lines from severe weather. And animals as diverse as elephants, prairie dogs and termites continually reconstruct the planet's crust, altering the flow of water, air and nutrients and improving the prospects of millions of species.[25]

Reduction in Greenhouse Gas Emissions

Distinct from the larger industrial model of beef production, 100% grass-fed beef offers profound potential for reducing greenhouse gasses.

METHANE

Many allegations about cattle's contribution to atmospheric methane may be true for conventionally raised cattle but are not true for 100% grass-fed beef cattle raised regeneratively.

Bovines raised with regenerative methods eat higher-quality forage than cattle raised conventionally. Green, leafy plants offer cattle better nutri-tion than corn and concentrates. Cattle digest higher-quality forages more quickly, reducing methane burps and lowering the amount of methane that the animal generates.[26]

Methanotrophic bacteria significantly reduce emissions because they live in pas-ture soil (among other places), and as cattle graze, these bacteria oxidize methane as their sole energy source.[27] Of course, this beneficial process does not occur where cattle are housed in a feedlot—or when cattle are removed from the pasture environment and enclosed in stainless steel rooms to measure their methane output. Also, tillage, nitrogen fertilizers, and bare land destroy methanotrophic bacteria.[28]

Scientists are also learning more about how water vapor transpired from pasture plants creates an oxidation zone whereby hydroxyl radicals break down methane.[29] Again, this process takes place in the *context of a pasture*, not in a feedlot or stainless steel box.

Because grass-fed beef cattle are not confined, they drop their manure all over the pasture. The manure and urine deposits are not concentrated and are processed by dung beetles. There are no manure piles or lagoons that release methane, as occurs in CAFOs.

NITROUS OXIDE

This potent greenhouse gas is formed and released in waterlogged areas where nitrogen fertilizer has been applied. By contrast, in a regenerative scenario mineral nitrogen is supplied (as nitrate or ammonium ions) to the plant by soil organisms as part of the natural system referenced at the beginning of this chapter. Farmers and ranchers should reduce and eventually eliminate the application of chemical fertilizers, including nitrogen, thus preventing the formation of nitrous oxide.

CARBON DIOXIDE

In addition to preventing the formation and release of carbon dioxide in the ways we've described, there are additional factors to consider when comparing the carbon footprints of corn-fed and grass-fed beef. The corn-based diet that fattens conventional cattle is dependent upon heavy fossil fuel inputs, notably from synthetic fertilizers, diesel-fueled tractors, and other equipment used in the production, processing, and storage of corn. In contrast, regenerative grazing makes good use of readily available resources: energy from the sun; low-tech tools that require little or no fossil fuel inputs; natural soil systems that can flourish on their own in perennial pasture; and the efficacy of cattle, pasture plants, microbes, dung beetles, and other organisms in the soil community that allow grassland ecosystems to function.

A 2018 report looking at greenhouse gas emissions associated with both feedlot finishing of livestock and regenerative pasture finishing notes that soil organic matter is 40 to 75 percent carbon, and soil erosion contributes to the release of carbon dioxide. The report notes further that such soil erosion on land used to produce cattle feed crops "should be incorporated in beef LCA [life-cycle analysis] accounting but has generally been excluded."[30]

Increased Carbon Storage

Efforts to reduce greenhouse gas emissions—while absolutely necessary—are not enough to address the climate change that is already under way. It is urgent that we pull carbon from the atmosphere and store it safely beneath the surface of the soil. In contrast to high-tech schemes such as engineering the clouds to refreeze the poles, or piping carbon dioxide to undersea storage units—two proposals that are under consideration—regenerative grazing fosters natural systems to sequester carbon, that is, systems such as photosynthesis and the activities of soil microbes.[31]

Studies that purport to show the climate footprint of beef tend to ignore the significant carbon sequestration that grass-fed beef production can achieve with regenerative grazing.[32] As noted, mycorrhizal fungi pass some of the carbon from plant roots into the soil, where it is stabilized in humus. Because of the ways that regenerative grazing safeguards soil microbes and enhances soil functioning—including carbon sequestration—the result is greater soil carbon concentrations than conventional grazing or no grazing.[33]

Multiple studies have shown that regenerative practices enhance, rather than inhibit, natural soil systems that sequester carbon, which is why 100% grass-fed beef operations can offer a net climate benefit, as expressed by Richard Teague in an interview:

> Our field research shows that even simple grazing, when you look after the grass reasonably well, will put more carbon in the ground than the emissions emitted by the cattle grazing—up to about three or four times as much. In the more sophisticated grazing systems we have been studying, there is an order of eight times as much carbon dioxide equivalents being sequestered into the soil as is being emitted by the cattle.[34]

Regenerative Grazing and Carbon Sequestration

The amount of carbon stored per year varies from place to place and year to year and depends on the condition of the soil initially. Eight metric tonnes per year (8.82 US tons per year) was reported by Megan B. Machmuller and colleagues in a study in the Southeast United States (Georgia) to determine how fast and how much soil carbon accumulates when soil that has been degraded by row cropping is converted to grazed pasture under intensive regenerative management. On the three farms in their study, they found a 75 percent increase in carbon stocks within six years of conversion to pasture.[35]

Has carbon storage by grazing been documented elsewhere in the United States?

Yes, carbon has been measured all over the United States. Most recently a large-scale study of carbon sequestration and storage brought about by regenerative grazing was conducted on ten farms in multiple states: Kentucky, Tennessee, Alabama (two pairs), and Mississippi.[36] The goal of the study was to compare the amount of carbon stored by regenerative versus conventional grazing. The study involved five "across the fence" pairs of livestock operations, with each pair having the same soil types and aspects. Each pair consisted of one operation with a history of AMP grazing (adaptive multi-paddock grazing with no fertilizer additions), and another operation with a history of continuous grazing (the conventional method with nonorganic nitrogen fertilizer additions). The study measured both carbon and nitrogen storage. Forty-two core samples were taken from each of the ten farms between May and June of 2018 and analyzed in a lab.

The results: the AMP grazing sites had on average 13 percent more soil carbon and 9 percent more soil nitrogen compared to the continuous grazing sites, in a 1-meter (1.09 yards) depth. These findings are evidence that AMP grazing management can be implemented at large scales as a way to store persistent carbon and mitigate rising atmospheric CO_2 levels in the atmosphere to combat climate change.

Is soil carbon permanent? How long does sequestration continue?

There is compelling evidence that the practice of regenerative grazing is a viable solution for removing carbon from the air and storing it for many years in the soil. Sequestering carbon in grass-lands is a dynamic process that depends on ongoing regenerative management of cattle on the land. Yes, soil microbes do consume carbon (and when those microbes die, their bodies contribute to soil fertility), but at the same time, other microbes are consuming root exudates and continuing to provide plants with nutrients in exchange for carbon, some of which will be eaten and some stored in the soil.

Richard Teague notes in an interview, "Where we have kept livestock on the land all the time under AMP management and not used fertilizers and pesticides, soil carbon is still going up after 15 years."[37]

Certainly, we must find ways of reducing climate emissions as well as storing carbon. But in light of the challenges countries have faced in reducing emissions, Dr. Rattan Lal, world-renowned soil scientist, in testimony to the US Senate Committee on Clean Air, Climate Change, and Nuclear Safety, described carbon sequestration as an essential strategy: "For the next 50 years . . . soil C sequestration is a very cost-effective option, a 'bridge to the future' that buys us time in which to develop those alternative energy options."[38]

Is there enough available land in the United States to keep large numbers of cattle on pasture for their entire lives?

Yes. Currently the United States raises about twenty-nine million corn-fed beef cattle per year, most of which graze on pasture until they get sent to a feedlot to be fattened on corn and other grains. Grass-fed beef expert Allen Williams added up the additional available acreage that would be needed for 100% grass-fed beef, including (1) land that now goes to growing grain for cattle feed, (2) unused or under-utilized land, and (3) acreage set aside in the CRP (Conservation Reserve Program), which pays farmers to take land out of crop production so that it establishes ground cover.[39] If US corn-fed cattle production switched entirely to 100% grass-fed production, we would have more than enough land to replace the corn-fed beef we're producing now with 100% grass-fed beef on a pound-for-pound basis.

From Theory to Practice

Many of the processes we've described in this chapter, such as storage of water and carbon, take place underground and may not grab our attention. This is not surprising; even soil scientists and ecologists have only recently become aware of some significant facts about grazing, such as the multiple ways that the pasture environment abates the release of methane to the Earth's atmosphere. So, because natural systems are so complex and work so well on their own, it may not be clear what farmers and ranchers can do to improve upon them! But there are plenty of things to do and other things to avoid doing, so let's summarize them here. The following are specific practices that will improve farmland health and productivity:

- Use adaptive multi-paddock grazing (AMP), with short grazing periods and long recovery periods for each paddock. (As we have noted, grass-fed beef production overseas typically does not use this method, and so it does not convey the full benefits to the environment that we describe in this book.)

- Keep soil covered year-round with growing vegetation (best option) or plant litter.
- Switch from plowing to no-till seeding in order to avoid damage to soil structure and soil microbes.
- Plant diverse cover crops (using no-till seeding) to extend the grazing season and to establish diverse microbial populations in the soil.
- Transition away from the use of biocides (pesticides and herbicides).
- Transition away from the use of chemical fertilizers, and manage for biological nitrogen-fixing instead. (For add-ons, favor organic amendments, targeted micronutrients, or formulations that feed soil microbial systems.)
- Use livestock bred to thrive in regeneratively managed grass-fed systems. (See the section "Genetics and Breeding" in chapter 6.)
- Convert marginal and degraded cropland to permanent pasture.
- Make transitions gradually in order to reduce risk, to learn from experimentation, to build confidence in new methods, and to allow time for populations of microbes to increase.

These are just basic guidelines. In part 2 we offer more details about regenerative husbandry, accessing wholesale markets, and making a profit—all while building soil health.

While more studies will be useful, we already have compelling data on the environmental advantages—and the net climate benefit—of regenerative grazing. In addition, we have the examples of people all around the United States who are successfully practicing this approach and realizing ecological goals as well as profitability on their farms and ranches. It's time for regenerative grazing of cattle to be universally celebrated and implemented as a clear and substantial strategy for a livable planet.

CHAPTER 4

The Roots of Health

In the end we all need to face the simple, bite-sized truth—
you are what your food ate.
　　　—DAVID R. MONTGOMERY and ANNE BIKLÉ[1]

Though you won't see it mentioned in the Dietary Guidelines for Americans, 100% grass-fed beef is a health food. A 2021 report, "Health-Promoting Phytonutrients Are Higher in Grass-Fed Meat and Milk," advances the idea that people who consume grass-fed meat or milk products benefit from the many pasture-plant nutrients concentrated in these foods.[2] This information adds to research going back several decades that documents health advantages of grass-fed beef over grain-fed beef: more omega-3 fatty acids, important for the brain and for protection against heart disease and arthritis; more CLAs (conjugated linoleic acids), which have shown promise in combating cancers and other diseases; and higher levels of vitamins A and E, both critical for health.[3]

And yet there is no consensus among nutrition experts about a place for grass-fed beef in dietary guidelines. It is widely acknowledged that differences in the diets of beef cattle are reflected in the nutrients and fats that humans get from eating the meat, but there is less agreement about the effects of those differences on human health. The absence of guidance on this important topic is due in part to three major problems with the existing body of research on beef consumption:

Most reports on the health impacts of eating beef do not distinguish feedlot cattle (fattened on corn) from grass-fed cattle (raised and fattened in pastures on grass and forage only).[4] Most do not even mention grass-fed beef, as if what cattle eat were of no health consequence for the animals or the people who eat the beef. But in fact, the research on "red meat" consumption has little or no bearing on health impacts of eating 100% grass-fed beef. Widely cited studies have conflated "red meat" with "processed meat." Processed meat is defined as meat preserved by smoking, salting, curing, or by the addition of chemical preservatives, to create products like bacon, sausages, hot dogs, salami, and ham, but the distinction between processed and unprocessed meat is blurred in research literature.[5] For example, one report refers to "scientific evidence of potential associations between unprocessed/processed red meat consumption and an increased risk of several chronic diseases."[6] The significance of this flawed methodology was revealed by a massive study that looked at consumption of processed meat and unprocessed meat separately. It found that while processed meat was associated with slightly higher risks of heart disease and diabetes, *no risk was found from eating unprocessed red meats.*[7] This analysis included twenty studies conducted in the United States, Europe, Asia, and Australia, which involved a total of more than 1.2 million individuals.

Many people eat red meat as part of a standard American diet—heavy on sugar, salt, and processed food—rather than a health-oriented, whole-foods diet that includes red meat.[8] In studies where no effort is made to account for the diets represented by the meat-eating cohort, the impacts of eating foods known to be harmful in quantity—such as sugar—are merged with the impacts of eating red meat.

While numerous studies have linked red meat consumption to chronic diseases—cancers, diabetes, and cardiovascular disease—recently this sweeping indictment of all red meat has been questioned.[9] For one thing, most nutrition studies are observational, which means that the conclusions are based on how well study participants remember and report what they have eaten, rather than what they actually ate. An international collaboration of researchers looked at the studies that have been cited as evidence for red meat's ill effects on health, and concluded that the science was weak. In 2019 they published a series of analyses in the *Annals of Internal Medicine* that found no compelling evidence that eating less red meat would benefit an individual's health.[10] These reports sparked a heated controversy, with the American Cancer Society and the American Heart Association sticking to their dietary recommendations, while other authorities welcomed the new perspective as being more science-based.[11]

Meanwhile, the myriad reports decrying red meat in recent decades apparently had an impact. In the United States, Americans consumed fifteen more pounds of beef per person in 1970 than they did thirty years later in 2000. In the same thirty-year period, they reallocated their meat eating toward poultry, accounting for 21 percent of their meat consumption in 1970 and jumping to 37 percent in 2000.[12]

But was this a boon for public health? Instead of a reduction in diseases, the US experienced an increase. By 2000, sixty-four million people in the US had "metabolic syndrome," a collection of risk factors that includes abdominal obesity, high blood pressure, and elevated blood sugar among others; fourteen million more people were diagnosed with these risk factors in 2000 than in 1990, despite a reduction in beef consumption during that period.[13]

The single fact that more people developed metabolic syndrome at the same time beef eating went down doesn't prove anything, but it does suggest a more complicated reality than recommendations to cut back on beef have indicated. Nicolette Hahn Niman, in her newly revised book, *Defending Beef*, marshals a sizable body of research that indicates a disconnect between red meat consumption and disease. Then Niman goes on to make a compelling case that the real culprit in our nation's ongoing health woes is not beef, but sugar.[14]

A scientist from the U.K. recently announced that the health of many young women is being compromised by a lack of vital nutrients because they consume little or no red meat and dairy products.[15] According to Ian Givens, director of the Institute for Food, Nutrition and Health at University of Reading, half of females between ages eleven and eighteen were consuming below the recommended levels of iron and magnesium and were also consuming too little iodine, calcium, and zinc.

Nutritional Density

Although we focus on the superiority of grass-fed beef over grain-fed beef throughout this book, we acknowledge that even feedlot beef offers significant nutritional benefits that may not be easy to get from a diet without meat. Whether grass-fed or grain-fed, red meat is nutrient dense. We will delineate specific advantages of grass-fed, but let's start by recognizing some undisputed benefits of red meat in general: high-quality protein, iron, zinc, and certain vitamins.

High-Quality Protein

Our bodies need protein to build muscle, transport nutrients, and build and repair tissue. "Complete protein" is defined as having all nine amino acids that are needed by the body and must be obtained through food. Most foods with complete protein are animal products, though plants also offer some protein, and soy offers complete protein.[16] (Vegetarians often combine foods that have complementary amino acid profiles with the goal of getting complete protein from a meal or within a day.)

Both animal and vegetable protein can be measured and scored for quality, that is, for digestibility. According to the two major standards used to score this quality (PDCAAS, protein digestibility-corrected amino acid score, and DIAAS, digestible indispensable amino acid score), foods from animals have higher protein quality than plant foods.[17] This is in part because plants have antinutrients (phytates and trypsin inhibitors) that interfere with digestion and absorption of protein.[18]

A global shortage of protein-rich foods is expected due to the COVID-19 pandemic and other factors, according to a 2020 report by the Food and Agriculture Organization (FAO). A decrease in protein consumption could exacerbate health problems among the poor, especially in children.[19]

Heme Iron

Iron has many functions; for example, it is involved in the transportation, use, and storage of oxygen in the body. Iron deficiency remains a big problem worldwide, especially in women and children. Women need larger stores of iron during pregnancy.

The dietary iron in animal foods, particularly in red meat, is heme iron, whereas the iron in plant foods, including grains, green leafy vegetables, pulses, and certain fruits, is non-heme. Heme iron is five- to tenfold more bioavailable than non-heme iron.[20] On the other hand, too much heme iron has been linked to cardiovascular disease, though this association has been found inconsistently. The background diet in which red meat is consumed may be a modulating factor; some studies find that the risk of too much heme disappears with diets that include whole plant food as well as red meat. This is because plant and animal foods can operate in symbiotic ways to improve human health, as in the example that follows of animal foods assisting with the uptake of zinc.[21]

Zinc

Zinc is an important trace mineral that is necessary for the creation of DNA, the growth of cells, building proteins, healing damaged tissue, and supporting a healthy immune system. Foods of animal origin have a high zinc content, while legumes contain very low levels of zinc.[22] People who restrict animal foods from their diet often have lower zinc status.[23] Uptake of zinc from plant sources can be lower as a result of the presence of antinutrients such as phytates, lectins, and certain fibers, but zinc uptake from plants can be improved when consumed in conjunction with animal foods.[24]

Vitamins

Vitamins A, B_{12}, D, and K_2 are more readily, or exclusively, obtained from animal sources as opposed to plant sources. These nutrients play essential roles in tissue development and regeneration, and some people are not able to get adequate amounts of these vitamins from plant foods alone.[25] Beef also has vitamins B_3 (niacin) and B_6 and other nutrients referred to as secondary that nevertheless play important roles in the body (for example, creatine, anserine, carnosine, and taurine).[26]

Pasture-Based Nutrition

Jumping into the controversy around meat eating, the 2021 report on health-promoting phytonutrients mentioned at the beginning of this chapter analyzed a growing body of evidence that indicates that grass-fed beef can offer nutrients that are only available for humans via meat (and milk) from grazing ruminants.[27] Unlike studies on beef that don't even mention what the cattle ate or where they ate it, this report is about meat from cattle and other ruminants raised and finished in pastures with a diversity of plants. The authors assert that the nutrition in the meat of pasture-fattened cattle can't be matched by corn-fed, feedlot beef. Furthermore, they state that the meat from pastured animals offers nutrients from plants

Bovine Self-Care

If you think a grazing cow chomps randomly on whatever she comes across as she lumbers around the pasture, think again. Cattle are very discerning when it comes to grasses and forage plants. They are able to snag precisely the plant they want—and not the neighboring plant—by wrapping their long tongues around their selection before tearing it off. Scientists say their discernment starts developing before birth, with exposure to flavors through amniotic fluid, and then following birth from tasting mother's milk, and finally from mom's grazing behavior in the pasture.[28]

I remember when I first grasped cattle's ability to know what their bodies need, not just on a regular basis but also in a remedial situation. I had created a supplement that I proudly labeled Ridge's Beef and Dairy Mineral Mix. It was a general chelated mineral mix to which I had added copper, iodine, and selenium (all naturally occurring in some—but not all—soils). The true test of my formulation came when I acquired a herd of cattle that had been under considerable stress, a situation that often can be addressed by making minerals available. So I put some of my mineral mix in the paddock where the new animals were grazing.

They barely touched it. But then I put out some kelp, and they gobbled that right up. So I guessed there was something in the kelp that they weren't getting in sufficient quantity from my mix.

To find out what that something was, I bought a cafeteria-style, free-choice, mineral dispenser from Advanced Biological Concepts, a company founded by the late Jim Helfter, who believed that most livestock health problems are due to nutritional deficiencies from single-source diets and related environmental conditions, such as feedlot confinement. Twelve of my new dispenser's slots were filled with various minerals. I filled the thirteenth with kelp and optimistically filled the fourteenth with Ridge's Beef and Dairy Mineral Mix. As soon as I put the dispenser in place, the cattle came over, nosed open the slots that held copper and iodine, and consumed substantial amounts of these two supplements, eating just a little kelp and leaving Ridge's Beef and Dairy Mineral Mix untouched. Though I continued to make my mix available even after this experience (okay, I'm stubborn), the cattle never showed any interest in it.

In hindsight, I don't know why I thought I could make a formulation to address all bovine needs under all conditions. Cattle may ingest a mixed mineral if something they need is not plentiful in the pasture where they are grazing, but a mix can also give them too much of something they do not need and therefore do not want.

—*Ridge*

that are not cultivated as vegetables for human diets, and therefore are not likely to be available in diets that exclude animal products.

According to their report, phytonutrients—the health-promoting, natural chemicals in plants—become concentrated in the meat and milk of grazing ruminants. The authors note, "Several phytochemicals found in grass-fed meat and milk are in quantities comparable to those found in plant foods known to have anti-inflammatory, anti-carcinogenic, and cardioprotective effects." They add, "Conversely, these phytonutrients are typically undetectable or present in lower concentrations in meat and milk from animals fed grain-based concentrates in confinement."[29]

The differences in nutrients found in grass-fed versus feedlot meat are not in total protein content, but in vitamins and trace minerals. For example, a comparison of riboflavin and thiamine in grass-finished and in grain-finished beef found nearly twofold higher riboflavin concentrations and threefold higher thiamine concentrations in grass-finished beef. Riboflavin is involved with the growth of cells, energy production, and the breakdown of fats, steroids, and medications; thiamine plays a critical role in energy metabolism and, therefore, in the growth, development, and function of cells.[30]

The report stresses that cattle (and sheep and goats) are healthier when offered pasture that has a wide variety of different grasses, forbs, and shrubs, as opposed to monoculture grasses. Such botanically diverse pastures improve the overall antioxidant content of beef, reducing the risk of diseases such as cancer and cardiovascular diseases. Also, diversity increases the likelihood of grazers finding the various primary and secondary nutrients that their cells need.

Van Vliet, Provenza, and Kronberg note, "While we often do not think of animals as intelligent beings, animals, unlike humans, do not have to be told what to eat, and nurture themselves prophylactically—to prevent disease—and medicinally—to treat disease."[31]

Seeing diverse pastures as "nutrition centers and pharmacies," the authors reference specific plant compounds (for example, phenols, terpenoids, antioxidants, and organic acids) that benefit the health of the grazing animals and the people who consume their meat or milk. By eating these food products, people can access nutrition that otherwise would be unavailable to them, largely because of the way our agricultural systems have developed:

Of roughly 200,000 species of wild plants on earth, only a few thousand are eaten by humans, just a few hundred of these have been domesticated, and only a dozen account for over 80 percent of the current annual production of all crops. . . . We narrowed the genetic basis of crop production to just a few plants, relatively productive in a broad range of environments, rather than broadening the range of plants that are valuable in local environments. We have also discovered only a fraction of the plant mixtures useful in nutrition and health and we have simplified agricultural systems in ways that are having alarming consequences on the health of people and the planet.[32]

The report cautions that more research is needed, but concludes that raising and finishing in pasture is beneficial to ruminant animals, and that these healthier animals provide healthier meat and milk.

More Grass-Fed Benefits

The new information about nutrients from pasture plants is in addition to a widely accepted list of nutritional advantages of grass-fed beef over grain-fed that includes more omega-3 fatty acids and CLAs (conjugated linoleic acid), and more of certain vitamins.

Fat and Fatty Acids

Fats are essential for our body's functions. For example, fat is needed to build cell membranes, the vital exterior of each cell, and the sheaths surrounding nerves.[33] During digestion, the body breaks down the fats we consume into fatty acids, which act in many different ways that influence health, well-being, and disease risk.[34] Stored fat moves certain vitamins (A, D, E, and K) throughout the body, and is the largest reserve of stored energy available for activity.[35]

While obesity worldwide has nearly tripled since 1975, fat is only stored in the body when we consume more calories than we use—*calories from any and all foods we eat, not just from dietary fats*.[36] For example, sugar in the diet can become stored fat in the body.

Leanness is typically listed as a health benefit of grass-fed beef, but this claim is misleading for two reasons.

First, fully finished grass-fed cattle are not lean in terms of intramuscular fat—the fat that contributes flavor, tenderness, and juiciness to cuts of meat. When grass-fed cattle are fully finished (fattened), their beef has as much flavor-packing intramuscular fat as grain-fed beef. It takes longer to fatten grass-fed cattle, and as bovines mature, the intramuscular fat is the last type of fat to be deposited.[37] Some producers unknowingly slaughter their cattle prematurely before they are fully fattened. Producers should learn how to examine their cattle for adequate fatness and wait until they are fully finished before sending them off.

In terms of the total fat, including the fat between the muscle meats and the external fat (under the skin), grain-fed carcasses do have more fat than grass-fed. But that extra fat is of low value in that it does not contribute to palatability, and most of it is removed, either when the carcass is broken into primal cuts, or when cuts for the meat case are closely trimmed. The removed fat is typically added to ground meat in a ratio specified by the wholesale buyer, which in turn is printed on packaged ground beef.

USDA Grading

Under USDA guidelines, beef carcasses are often graded for both yield and quality:[38]

Yield grading provides an estimate of the percentage of boneless, closely trimmed retail cuts of meat.

Quality grading evaluates palatability: flavor, tenderness, and juiciness.

The flavor is in the fat, which also has some influence on tenderness and juiciness—the whole "eating experience."[39] Marbling is the mingling or dispersion of fat (intramuscular fat) within the lean. The degree of marbling is the primary determinant of the quality grade. The degree of maturity (referring to physiological rather than chronological age) may also be a factor.

USDA meat graders are trained to make a visual appraisal of the fat density by cutting the carcass at the thirteenth rib and assigning a marbling score for each beef carcass. The more flecks of white that the grader sees, the higher the grade. Since the great majority of carcasses seen by USDA graders are of feedlot cattle, they are accustomed to seeing and judging the very visible marbling from corn-feeding as evidence of intramuscular fat.

There are three grades above standard: select, choice, and prime, with prime being the highest grade. The added value of a prime carcass typically benefits the feedlot or packer rather than the rancher.

Even more important, while westernized cultures see leanness as a positive attribute, dietary fat is actually critical to health, and the fat from grass-fed cattle offers health benefits not found in grain-fed beef. While hunter-gatherers such as the Maasai in southeastern Africa—whose diets are high in milk, red meat, and fat—have less heart disease and cancer than people who eat a diet high in processed foods, Americans remain leery of animal fats.[40] This is hardly surprising: For many years health professionals worldwide have recommended reducing consumption of fatty meat and whole milk, certain that these foods would lead to hardening of the arteries and heart disease.[41] As a result, in recent decades many people have cut back on dietary fats significantly, including most animal proteins, such as meat and dairy products. And yet heart disease world-wide has increased alarmingly, according to a 2020 release from the American College of Cardiology that calls for emergency measures through prevention programs and access to emergency care and medications.[42]

Similarly, some formal studies attempting to pinpoint a connection between eating animal fat and disease have found an inverse relationship: more consumption of animal fat correlating with less disease.[43] For example, in regard to whole-fat milk, an eighteen-year study that included over twenty-five thousand individuals showed that children who drank whole milk had a lower risk of becoming obese as compared to children who were given fat-free or 1-percent fat milk. A number of studies have demonstrated associations between more full-fat dairy in the diet and reduced risks of type-2 diabetes and cardiovascular disease.

A 2020 report notes that a growing body of literature has found that two types of saturated fatty acids (even-chain and odd-chain fatty acids) have opposite associations with disease and health: one associated with low risk for disease and the other with high risk.[44] And yet dietary guidelines that call for reducing saturated fat make no distinction between these two types, which has led to calls "to refine global saturated fatty acid dietary guidelines."

Similarly, Cynthia Daley, professor in the College of Agriculture at California State University, has noted that lipid research suggests that not all saturated fatty acids have the same impact on cholesterol, and that fatty acids should be consid-ered individually when making dietary recommendations for the prevention of cardiovascular disease.[45] Cholesterol guidelines were also called into question in 2009 by a national study that was based on data from 136,905 patients hospital-ized for a heart attack; the study found that nearly 75 percent of those patients were not at risk for a heart attack based on cholesterol guidelines.[46]

And yet dietary recommendations from the American Heart Association, 2021, and the USDA Dietary Guidelines, 2020–2025, recommend reducing or eliminating fats.[47] One of many studies that casts doubt on those guidelines is a September 2021 report on a randomized controlled trial that followed over four thousand people for more than sixteen years and concluded that dairy fat does not adversely affect either cholesterol or blood pressure, which are seen as biomarkers for cardiovascular disease.[48]

ESSENTIAL FATTY ACIDS AND OMEGA-3S

Science recognizes two distinct families of fatty acids that are termed "essential," because they are critical for human health but the body cannot synthesize them and they must be obtained from food. Omega-3 fatty acids are one essential family; the other essentials are omega-6 fatty acids.

Both omega-3s and omega-6s are polyunsaturated fats that have important functions in the body, particularly relating to inflammation and blood clotting. But because the two essential fatty acids are from different families, excess of one fatty acid family can interfere with the metabolism of the other.[49] Therefore, nutrition experts agree that it is important that the ratio of one family to another should be balanced—or close to a balance—in human diets.

Nutrition experts also agree that in Western diets the current ratio of the two essential fatty acids is very much out of balance, with way too much omega-6 relative to omega-3. The recommended ratio of omega-6 to omega-3 in a healthy diet varies in the range of 1:1 up to 4:1, but in the typical American diet, the range is estimated to be from 15:1 to 16.7:1.[50] One reason suggested for this imbalance is that our diets are high in processed foods. Whereas omega-3 is anti-inflammatory, the imbalance with an excess of omega-6 may explain increased incidences of heart disease, cancer, rheumatoid arthritis, autoimmune disorders, and neurodegenerative diseases thought to stem from inflammation.[51]

While fish have more omega-3s than beef, grass-fed beef has significantly higher levels of omega-3s than grain-fed beef. And a research review shows that as the concentration of grain *increases* in the bovine's diet, the concentration of omega-3 fatty acids *decreases* in a linear fashion.[52] Grass-finished beef consistently produces a higher concentration of omega-3s without affecting the omega-6 content, resulting in a more favorable omega-6 to omega-3 ratio.

In regard to increasing the level of omega-3 fatty acids in our bodies, a 2012, randomized, double-blind, dietary intervention report offers an example of how eating pasture-finished meat can affect our blood levels of omega-3s.[53] The study was carried out for four weeks on thirty-eight healthy subjects (males and females) divided into two groups, which ate a given amount of grass-finished or grain-finished meat respectively, every week for four weeks, with both groups avoiding consumption of oily fish during the test period. Based on weight, blood pressure, and blood samples, all taken as a baseline and again at the end of the intervention, the report concluded that eating grass-finished meat had increased the blood levels of two essential omega-3 fatty acids, EPA and DHA, both of which are considered to be important in the prevention or treatment of diseases, and are also essential for fetal development and healthy aging; EPA fatty acid reduces inflammation and depression; DHA fatty acid is found in abundance in the brain and retina.[54]

CLAS (CONJUGATED LINOLEIC ACIDS)

CLAs are a group of fatty acids that enter the human body primarily from ruminant animal products: meat and milk. Numerous reports have indicated that CLAs are beneficial to human health, as demonstrated by animal models.[55] Researchers have found promising—though not definitive—evidence that CLAs can combat obesity, cancers, hardening of the arteries, diabetes, and other diseases.[56] But to date, scientists have not fully determined (1) the mechanisms involved, (2) to what extent CLAs are effective, and (3) against which diseases.

CLAs are formed by a bacterium in the animal's rumen (that is, the first compartment of the four-part digestive system), where cellulose is broken down. A diet of green forages increases the concentration of CLAs in the animal's fat; the CLA content is 300 to 500 percent higher in beef and dairy from grass-fed cows than in grain-fed cows.[57]

Although CLAs are technically trans-fatty acids because of their chemical structure, the US Food and Drug Administration (FDA) exempts CLAs from classification as trans-fat on nutrition labels because they are naturally occurring components of food that are not associated with coronary heart disease risk factors.[58]

Initial positive reports on disease-fighting benefits of CLAs were based on cell culture models and animal studies, which were followed by studies involving humans.[59] The results of the studies involving humans have been variable. For example, several early studies indicated that CLAs could be effective in fighting breast cancer and colon cancer.[60] In regard to breast cancer, women who consumed *lower* levels of CLA had a 3.3-fold *greater* risk

Grass-Finished and Fat

I was already yawning as I settled into my chair to hear an after-lunch conference presentation. But the speaker, Loren Cordain, author of *The Paleo Diet*, snapped me to attention with his description of Paleo people, armed only with sharp wooden sticks, hunting massive, long-horned aurochs. These now-extinct bovines stood six feet tall (modern cattle are four to five feet tall) and weighed as much as 3,300 pounds. Because these wild herbivores ate no grain, some people might call their meat "grass-fed beef." Cordain speculated that the Paleo people undertook these perilous aurochs hunts to get lean meat.

Grass-fed bovines are often characterized as lean—not only the aurochs but also their descendants: today's grass-fed cattle. But is it true that they are leaner than their grain-fed counterparts?

Our experience supports the conclusion of the Stone Barns Center for Food and Agriculture report, *Back to Grass*, that grass-fed beef is not lean if it is well finished.[63] When we were starting out with grass-fed beef, we took one of our carcasses to butcher Mike Debach of Leona Meat Plant near Troy, Pennsylvania. When Mike butchered the carcass, he didn't see much marbling

of breast cancer than women who consumed higher levels. Regarding colon cancer, "total high-fat dairy food consumption was significantly and inversely associated with the risk of colorectal cancer"; in other words, the more dairy the less cancer.[61] But subsequent studies involving humans, performed by different groups with different study cohorts, did not confirm the positive results of the earlier studies.[62]

A 2011 review of studies of both CLAs from milk and CLAs from supplements noted that "most human intervention studies have utilised synthetic so he told me he didn't expect the meat to be tasty. I said, "Do me a favor. Take a steak home, cook it, and taste it." Mike prepared the steak that very night and was astonished at the rich flavor. The next day, with amazement in his voice, he reported, "Ridge, that was one of the best steaks I ever had!"

This led me to believe that my grass-fed beef had plenty of intramuscular fat, though the marbling was not easy to see. To settle the question, we took eight of our grass-fed steaks to Dr. Susan Duckett at Clemson University to grade the meat for quality. Instead of using a visual appraisal—which is the way meat is graded by the USDA—Duckett precipitated the fat to make an accurate measurement. The results: not one steak was lean. All eight of these 100% grass-fed steaks were fat and most were very fat; one was prime, six were choice, and one was select.[64]

We are not the only ones to notice that the intramuscular fat of grass-fed cattle is not always as visible as the fat of grain-fed cattle—but a taste will tell you that it's there![65] As to those Paleo people hunting aurochs, we can't speak to the palatability of aurochs meat, but we can attest that fully finished 100% grass-fed beef is tender, juicy, and flavorful.

—Ridge

CLA supplements."[66] This may be because early reports of CLAs' effectiveness against disease created much excitement, which spawned an industry of CLA supplements that in turn prompted studies about their makeup and dosage.

A 2021 comprehensive review of the research literature on CLAs concludes that overall, "These naturally occurring substances have demonstrated to have anti-carcinogenic activity," and suggests that differences in results from study to study may be in part due to the difference in dosage of CLA uptake.[67] Optimal dietary intake remains to be established for CLA.[68]

Vitamins and Antioxidants in Grass-Fed Beef

The superior nutrition goes beyond high-quality fat and protein—100% grass-fed beef is also chock-full of vitamins and antioxidants.[69]

VITAMIN A. The muscle meat of pasture-raised cattle has seven times higher beta-carotene levels than grain-fed beef. The yellow tint of the fat in grass-fed beef comes from the green forages that the cattle have eaten; the foliage has carotenoids (mainly beta-carotene), which are precursors of retinol (vitamin A), a critical fat-soluble vitamin that is important for normal vision, bone growth, reproduction, cell division, and cell differentiation.

VITAMIN E. Vitamin E is beneficial to vision, reproduction, blood, brain, and skin. Cattle finished on pasture have three times more of this powerful antioxidant than cattle finished on high-concentrate diets such as feedlot rations.

ANTIOXIDANTS. Some of the vitamins in grass-fed beef are antioxidants, which reduce risk of disease by counteracting free radicals, which contribute to chronic diseases such as cancer and cardiovascular disease. Note that the action of antioxidants is chemical, rather than nutritive. Because free radicals lack a full complement of electrons, they steal electrons from other molecules and damage those molecules in the process; antioxidants neutralize free radicals by giving up some of their own electrons. Pasture raising improves the overall antioxidant content of beef.

Preliminary research shows that vitamin E, an antioxidant, may help prevent heart disease. It may also block the formation of carcinogens from nitrates consumed in the diet, and may also protect against cancers by enhancing immune function. Glutathione (GT) is a newly identified protein antioxidant that can quench free radicals within a cell. Grass-fed beef is high in GT because of the green forages the cattle consume.

Regarding all nutrients, the 2020 report "Plant-Based Meats, Human Health, and Climate Change" suggests that the effects of synthetic supplements are different from the effects of nutrients obtained from a natural diet.[70] Whole foods are extremely complex, and the authors point out that ingesting a single nutrient outside of its natural food matrix may not provide the desired health benefit. Research studies have shown how different the results can be, depending on whether nutrients were ingested as supplements or in foods; and the 2020 report offers examples of various outcomes:

- Adequate intakes of zinc, copper, and vitamins A and D were associated with decreased risk of cardiovascular disease and all-cause mortality when obtained from foods, but not from supplements, in a recent large population-based study.
- Carotenoid-containing foods are associated with a decreased risk of various cancers and cardiovascular disease. However, studies suggest that carotenoid and/or vitamin A supplements do not decrease the risk of cancer or cardiovascular disease.
- Studies show a potential for increased cardiovascular disease risk with calcium supplements, but not when calcium is obtained from food.

Threats Avoided

In addition to direct health benefits of eating grass-fed beef, some serious risks are reduced or avoided by choosing grass-fed over grain-fed beef, including the following health threats: antibiotic-resistant bacteria, E. coli, mad cow disease, and nutrient deficiency.

Antibiotic-Resistant Bacteria

One of the increasing dangers of feedlot practices is the development of antibiotic-resistant bacteria. In 2015 the American Academy of Pediatrics issued a warning that the resistance of microbes to antibiotics is "one of the most serious threats to public health globally and threatens our ability to treat infectious diseases."[71] The release cautioned parents that children can be exposed to multiple-drug resistant bacteria through consuming the meat of animals dosed with an antibiotic.[72]

Heavy dosing of livestock with antibiotics has brought about the development of bacteria that cause illness but do not respond to formerly lifesaving medication. In 2019 the Centers for Disease Control and Prevention (CDC) reported that more than 2.8 million infections that do not respond to treatment occur each year, with more than thirty-five thousand resulting deaths.[73]

Feedlot life is stressful for cattle. Because the corn-based diet and the concentration of penned animals lead to numerous health problems, feedlot

A Sugar-Coated Fact

Everyone has heard about grass-fed beef and corn-fed beef—but *candy-fed*?

A few years ago, a truck full of Skittles "seconds" spilled part of its load onto a highway. The candy was headed to a Wisconsin farm where it would be part of the cattle's daily ration. The fact that candy was being fed to cattle was covered by the *Huffington Post* and caused an uproar in some quarters.

Actually, feeding discarded candy, baked goods, and factory-prepared sweets to livestock has occurred for years and has been commonplace since 2012. Corn prices surged that year, driving up the cost of corn-based feeds, so beef producers were eager to find a substitute carbohydrate. At the same time, food manufacturers were trying to avoid the expense and environmental harm of dumping their waste products in landfills. (Food waste in landfills produces methane.) Thus, the candy-to-cattle practice took hold.

To this day, some farmers and nutritionists serving the cattle industry—but not the grass-fed cattle industry!—argue that the

cattle are routinely given antibiotics along with their food to keep them from contracting a fatal disease before they are fattened and slaughtered. Antibiotics are also given to promote rapid growth.[76] For these purposes, feedlots typically add "subtherapeutic" medication to the feed or water of an entire herd to promote faster growth with less feed. This misuse of antibiotics is linked to the development of human pathogens that are resistant to antibiotics.

Some beef producers allow antibiotics to be used in the case of illness but prohibit antibiotics fed to cattle as a precaution or to speed growth. Other meat companies (including ours), which brand their meat "antibiotic free," have a producer protocol that prohibits antibiotics for any reason, either in sugar in candy is a fine substitute for corn carbohydrates; it is economical for beef producers and avoids wasting candy that is being discarded. Federal regulations? No problem. That rule book focuses almost entirely on animal safety and foodborne diseases that affect humans.[74] The health effects of humans eating the meat of cattle that regularly consume candy apparently has not been studied.

Other allowable feeds for confined cattle include feather meal and poultry litter. Again, this is nothing new; a research report from the University of Florida dated 1994 says, "When fed to yearling cattle, liquid feed containing feather meal produced faster gains and higher pregnancy rates than liquid feeds containing urea. When fed to yearling or breeding female cattle up to four years of age, feather meal should be the crude protein of choice. Feather meal has a far superior nutritive value to urea and will produce favorable economic returns."[75]

feed or by injection. This kind of prohibition in a protocol generally means that if a farmer chooses to treat an animal with an antibiotic, that animal would have to be sold elsewhere.

Escherichia coli (E. coli)

Another threat from feedlot management is E. coli, a bacterium found in the large intestines of warm-blooded animals, including healthy humans and cattle. Most strains of this bacteria are harmless or cause little harm, but a few strains can cause serious illness, even death, especially in children or elderly people. The goal is to protect people and animals from these strains, which can be transmitted through food.[77]

The corn-based feedlot diet creates an acidic condition in the bovine rumen that allows acid-resistant *E. coli* to develop. If passed on, the acid-resistant *E. coli* can also survive the acid of the human stomach and cause illness, paralysis, or death.

Cattle evolved digesting roughage that is fermented slowly in the rumen. The rumen microbes of forage-fed animal are selected for fiber digestion. Adjusting cattle to high-grain diets from predominantly forage diets disrupts the normal microbial environment and causes acidosis. Beef producers can reduce the chances of causing or spreading *E. coli* infection by feeding cattle a natural diet of grass or hay.[78]

The effectiveness of the natural diet against *E. coli* was illustrated by a Cornell University study that found that switching cattle from grain to hay for the last two weeks of their lives reduced the incidence of *E. coli* significantly.[79] Cattle raised entirely on pasture don't suffer from stomach acid, which means that grass-fed cattle are unlikely to pass on *E. coli* to humans.

Mad Cow (Spongiform Encephalopathy)

This fatal disease, caused when people or cattle eat infected beef products, was first identified in Britain in the 1980s and has been transmitted to humans in the United States and abroad.

A case of mad cow in the US in 2003 made headlines and caused a loss of some foreign markets for several years. Subsequently, FDA regulations have prohibited serving ruminant protein to ruminants. Since then, there have been only "atypical" cases of mad cow—the most recent being in Florida in 2018—and these cattle did not enter the food supply. Consumers can avoid mad cow disease by eating beef raised on a diet that consists entirely of grass and forage.[80]

Nutrient-Deficient Food

Between 1940 and 1991, scientists in the U.K. documented a decline in the nutritional value of a number of foods, including vegetables, fruit, dairy products, and meat.[81] The items evaluated were not highly processed foods, but an array of whole foods such as carrots, oranges, beef, and other dietary staples. Over fifty-one years of data collection tracked a significant loss of minerals and trace elements in food. Similar findings have been found in the US.[82] Some scientists find this decline unsurprising given the depletion of nutrients in our soil.[83] Poor soil leads to nutrient-deficient food on the table.[84]

Healing Agriculture?

In an interview for *Acres U.S.A.* magazine, Australian soil ecologist and bio-chemist Christine Jones talks about the connection between nutritious food and healthy soil. She stresses the critical role of soil microbes as a bridge between soil and plants. Jones speaks of the current threat to human health as a "double whammy": not just the toxins in our food, but also the use of agricultural chemicals that kill soil microbes.[85]

These microbes, particularly the soil fungi that we described in chapter 3, must be allowed to flourish because they bring water, essential nutrients, and trace elements to growing plants, including vegetables, grasses, and other forage plants. Trace elements are necessary for the plants to make the phytonutrients that we need to combat disease. As Christine Jones said in the *Acres U.S.A.* interview:

> If the plant-microbe bridge has been blown, it's not possible for us to obtain the trace elements our bodies need in order to prevent cancer—and a range of other metabolic disorders. Cancer is not a transmissible disease. It's simply the inability of our bodies to prevent abnormal cells from replicating. To date, the response to the cancer crisis has revolved around constructing more oncology units, employing more oncologists and undertaking more research. The big breakthrough in cancer prevention will be in changing the way we produce our food.[86]

Similarly, in the report on the health benefits of grass-fed meat and dairy, authors Provenza, Kronberg, and Gregorini assert that "more money and effort ought to be spent creating human and environmental health by growing and eating wholesome foods and less effort spent treating symptoms of diet-related diseases."[87]

Agricultural systems that include regenerative grazing on a large scale will go a long way toward providing healthy, nutrient-dense food for growing populations. Part 2 of this book delineates specific practices in use today on some farms and ranches to bring about these results.

CHAPTER 5

Animal Welfare

We can say that a cow grazing on grass is at least doing what he has been splendidly molded by evolution to do. Which isn't a bad definition of animal happiness.

—MICHAEL POLLAN[1]

In a book about raising cattle to be killed for food, what does *animal welfare* mean? For many of us who eat meat, animal welfare means that animals raised for food should be free from suffering throughout their lives, and that their deaths, too, should be free from pain or fear.

The American Humane organization has developed criteria for livestock welfare as the basis for their US farm animal welfare certification program, which they say now covers more than a billion animals.[2] They call their criteria the Five Freedoms:

1. freedom from hunger and thirst
2. freedom from discomfort
3. freedom from pain, injury, and disease
4. freedom to express normal and natural behavior
   (e.g., accommodating for a chicken's instinct to roost)
5. freedom from fear and distress

Livestock raised with humane practices can experience their lives and deaths without losing any of the Five Freedoms. It may be useful to consider the experiences of well-treated domesticated animals in contrast to the experiences of animals in the wild. People who spend a lot of time in the natural world are aware that wild animals are not free from hunger, thirst, discomfort, pain, injury, disease, fear, and distress; they are likely to experience any or all of these states in the course of their lives. And because life lives off of life, their deaths are often painful and protracted, as when they are ripped open by predators and eaten while still alive.

Increasingly, some of the suffering and deaths of wild animals—for example, rabbits, worms, turkeys, butterflies—are a direct result of humans growing crops, including vegetables and grains. Determining how many wild animals are killed under these circumstances depends on how the question is posed. For example, does the number include deaths that are not caused by machinery or agricultural poisons but by farming activities that destroy habitat? Examples of the latter include the loss of monarch butterflies in Mexico, where their habitat continues to be cleared for avocado plantations;[3] the loss of 82 percent of orangutans in Borneo, where their habitat has been cleared for oil palm plantations;[4] and the loss of 50 percent of grassland birds in North America in the last fifty years from destruction of their habitat due to large-scale farming.[5] Another question: Are some animal lives more significant than others? If insects are to be counted in the death toll from farming activities, the true number of deaths are probably incalculable.

An estimate of the number of unintentional deaths caused by agriculture in the United States that includes only mammals, fish, reptiles, and other amphibious creatures ranges from 63.75 million to approximately 127.5 million per year.[6] For comparison, 33 million cattle were slaughtered in the United States in 2019.[7]

While there is no question that many animals raised for food are mistreated or neglected, this is not always the case and is certainly not necessary. Improvements can and should be made in both livestock and vegetable production to improve the well-being of animals, both wild and domesticated, though it should be acknowledged that not all suffering and death can be eliminated.

Context for Care: Pasture or Feedlot?

Since 97 percent of beef in the United States comes from cattle fattened in feedlots, let's compare the living conditions of cattle that spend their entire lives in a pasture to the conventional scenario, whereby cattle start their lives on pasture but spend their last four months on a feedlot.[8]

Living Conditions: Pasture

We see the Five Freedoms as integral to raising 100% grass-fed beef cattle the way we've described in previous chapters. Regenerative management practices provide the best foundation for the health and well-being of cattle. These practices are shaped by the cattle's needs as ruminants and herding animals.

Grass-fed cattle are born, raised, and fattened for market on pasture and are never confined in a feedlot of any kind. They graze on a natural diet of grasses and other forage plants, supplemented by hay in winter as needed, and they are generally healthy, requiring little or no medical care. On Big Picture Beef's partner farms, supplemental minerals may be offered, but no grain, no industrial by-products, no growth hormones, and no antibiotics are given.

The best grazing and pasture management method is adaptive multi-paddock grazing, as detailed in chapters 3 and 6. The cattle spend most of their time grazing and the rest of their time resting or socializing in small groups of their own choosing. They are allowed to care for their young as their instincts as bovines dictate.

The cows in our Big Picture Beef program are bred to calve in the spring or the fall. Calves are born in the pasture, typically in a protected place chosen by the mother. With careful breeding, proper nutrition, and excellent health, they almost never require assistance with birthing, and the calves thrive under their mother's care. Our protocol requires that calves be left with their mothers to nurse for a minimum of six months, with ten months preferred.

In the Northeast, pastures are lush from April to October, but given a choice, these rugged animals prefer to be outdoors even in winter, when the grass has stopped growing and is often covered in snow. Then they either graze on stockpiled grass or eat hay. They are naturally protected from harsh weather by a thick layer of fat, a furry winter coat, and a rumen (the fourth part of the bovine digestive system), which ferments at about 104 degrees Fahrenheit and serves as a self-heating system.[9] After our first winter of caring for a herd of Rotokawa Devon cattle in a New England pasture, with some sheltered areas but no barn, we had the veterinarian come in to evaluate the animals. He said that they were the healthiest cattle that he had seen that spring, all the others having spent the winter enclosed in barns.

Cattle are herding animals and express discomfort if they are temporarily isolated from the herd. They also recognize the people who look after them. Because the breeds that we raise and recommend have a docile temperament, farm families usually handle them with minimal effort. When it's time for the herd to move to a new paddock for fresh grass, the grazier moves the flexible fence and the animals walk into the new area readily, with the young ones frisking in the tall grass.

To ensure that the farms in our Big Picture Beef program follow our methodology and specific requirements, we require our Northeast farm partners to sign our protocol, which is detailed in chapter 6. We describe more good husbandry practices in chapters 6 and 7.

Not all grass-fed beef operations are managed the way we describe, but this approach is embraced and practiced by the grass-fed beef producers we know, and not just in the Northeast; the protocol of Thousand Hills Cattle Company in Minnesota is almost identical to ours.[10] The Regenerative Organic Alliance, founded in 2017, has adopted certification criteria that include the Five Freedoms, which is another example of regenerative farming being associated with humane care and handling of livestock.

Living Conditions: Feedlot

In contrast to 100% grass-fed beef cattle, conventional beef cattle are raised according to a very different concept of animal welfare. In the beef industry, profits are prioritized over natural systems and animal instincts, as illustrated by this quote from the Penn State Extension website, in which concerns over drugs fed to cattle are described as "setbacks":

> Technology will drive the improvement of beef production and keeping food prices low now and in the future. Setbacks in the use of feed additives, implants, drugs, and other technology will slow that improvement. However, customers require a positive perception of the food they buy, and it is necessary to effectively answer their concerns.[11]

Regardless of where in the United States they were born, beef cattle are typically trucked to a feedlot at ten to twelve months of age, where they are fattened for four months before being slaughtered. Forty percent of feedlot

THE LEARNING CURVE

Mother Cows Know Best

On Roger Fortin's Little Alaska Farm in Maine, he had a mix of British breeds (Hereford, Angus, and Devon) with the moderate frames and deep bodies that are characteristic of cattle that thrive in a grass-fed beef program.

Roger usually weaned his calves in the fall at five or six months of age, following the custom of other livestock farmers in his area, but he was pretty sure that his rugged mother cows could nurse their spring calves through a winter, while maintaining their own excellent condition. So he told me about a trial that he was planning, and it sounded good to me.

As a former dairy farmer, he had an open free-stall barn and a bunk silo, which was the setting for the experiment. In the fall he selected six of his May calves to be weaned, and six others to stay with their mothers through the winter. While the two groups had separate pens for the winter, all the calves would receive the same ration of grass silage at the same bunk in the same open barn.

That winter, the only difference in the experience of the twelve calves was that six of them were spared the stress of weaning, and instead had the comfort of their mothers' companionship plus

cattle are at facilities housing thirty-two thousand head or more.[12] Instead of grazing on pasture, they are confined to pens with no grass, each pen typically housing 100 to 125 animals, where the cattle stand in mud and manure. Their corn-based rations, unnatural for ruminant animals, are brought to the pens. The diet, the confinement, and the lack of opportunity to graze are inherently stressful for cattle. Antibiotics are routinely given in food or water to prevent or treat disease, which occurs often as result of feedlot conditions, especially the corn-based diet.[13]

some mother's milk, which by the end of winter had dwindled to perhaps a half cup per day. In March, when the nursing calves had essentially self-weaned at ten months, the six nursing mothers were still in excellent condition.

Roger weighed all twelve calves that were in the experiment, and the six nursing calves were on average seventy-five pounds heavier than the six calves that had been weaned before winter. Amazed, Roger asked me, "Do you know what this means?"

What it meant was that the six calves that had continued to nurse could go to market in the fall. Because they were going into the grazing season seventy-five pounds ahead of the rest, with well-managed grazing over the spring and summer, the chances were good that those calves would be fully finished and ready for slaughter at eighteen to twenty months of age. The other six would have to be maintained on stored feed for a second winter at the cost of about $2.75 per calf, per day, every day from October into April.

In a business with small profit margins, every cost-cutting measure is significant. And in grass-fed beef production, the practices that are best for the cattle's health and well-being are often the best way to achieve net profitability, as well.

—Ridge

"Standard" Abuse?

Feedlot conditions systematically deprive cattle of some of the Five Freedoms advocated by the American Humane organization. Keeping cattle confined to grassless pens deprives them of the freedom to express their most characteristic natural behavior: grazing.

Also, the economics of feedlot beef production dictate that cattle must gain weight at their maximum potential rate; this involves getting them quickly onto a diet containing a high concentration of grain, which can lead to grain overload.[14] This diet causes health problems such as bloat and liver abscesses, which result in extreme discomfort and sometimes death. (The incidence of liver abscesses in slaughtered beef cattle average from 12 to 32 percent in most feedlots.[15]) Another ubiquitous health issue on feedlots is acidosis, a digestive disorder caused by diets that are high in grain and low in roughage. But these practices—confinement and corn-feeding—are considered "standard" rather than abusive treatment.

Regarding commonly used drugs, most are given to cattle for two reasons: (1) to correct a condition that would not exist if the animal's natural needs were being addressed, or (2) specifically to increase profits.

In recent years the medical community has focused attention on the dangerous feedlot practice of administering antibiotics to prevent or treat health problems that are caused by feedlot living conditions, as we described in chapter 4. Conventionally raised cattle that are not old enough for the feedlot may be given antibiotics for conditions created by inappropriate husbandry practices on the farm or ranch where they are raised. For example, the following quote from Beef2Live, a website addressed to beef producers, describes the common practice of giving antibiotics to calves when they become ill due to abrupt weaning or because their rations have a high percentage of corn:

> The period around weaning, or when calves are separated from their mothers, is a time in which antibiotics are used more commonly than other times. The stress of separation, sometimes along with very variable weather conditions, can make calves more vulnerable to pneumonia. . . . Since this illness, if it occurs, occurs in the first 2–3 weeks after the stress, these uses of antibiotics are short-term, focused around that time period, and fed in higher concentrations. . . . Another use of feed antibiotics may occur when calves are growing rapidly and approaching their final weight. Calves can develop abscesses (pockets of infection) in their livers when a high percentage of their diet consists of grain. In some cases these abscesses can cause illness in the calf. They result in the liver not being fit for food. To prevent this condition, antibiotics such as tylosin are fed to cattle late in the feeding period.[16]

Regarding drugs given for the sole purpose of increasing profit, beta-agonists are administered to increase sales price rather than to treat illness.

These drugs, which are given to an estimated 70 percent of cattle, bulk up muscle, making the cattle carcasses more saleable. But the drugs also make the live animals prone to lameness at the end of their lives.[17] In a 2021 paper, Dr. Temple Grandin, from Colorado State University, explained how this impairment prompts slaughter facility employees to use electric prods to get the cattle to move faster; she cited the use of beta-agonists as one of the causes of lameness.[18] This class of drugs has been in the news off and on since 2013, when Tyson, one of the "Big Four" meat processing companies, refused to buy cattle that had been treated with Zilmax, a beta-agonist, because some of these animals were arriving at the Tyson's packing plants unable to walk.[19] But beta-agonists are still in use.

Humane Handling of Cattle

In learning how best to care for the well-being of cattle throughout their lives and at the time of their deaths, we have been guided by the work of Temple Grandin, who has designed livestock handling facilities all over the world and has developed practices that reduce or eliminate animal stress and pain in connection with livestock management.

For example, regarding weaning calves, Grandin recommends the lowest-stress method, which is to let a calf wean itself. If that is not successful, the next best method is one that keeps the cow and calf in close contact, such as fence-line weaning, where the pair are penned separately from each other but can remain in contact through the fence. She notes that cows and calves are more concerned about remaining close than they are about nursing.

Grandin points out that a calm, caring approach to handling livestock is not only humane, it is also good business. Caring for cattle in a way that makes them content fosters good health; the animals will have increased weight gain, better reproduction, and fewer bruises or injuries. Her book *Guide to Working with Farm Animals* has many handling tips and is illustrated with photos and diagrams. Most of the following advice is adapted from this book, which we strongly recommend for all livestock producers.

How Cattle Remember Fear and Pain

Animals store fearful experiences as sensory memories: what the animal saw, heard, smelled, touched, or tasted when the frightening or painful event

Temple Grandin

Known worldwide as an expert on animal handling, Temple Grandin feels that being a high-functioning autistic person has given her insight into the sensitivities and behavior of cattle and other livestock.

She received her PhD in Animal Science in 1989 and currently teaches courses on livestock behavior and facility design at Colorado State University. She also consults with the livestock industry all over the world. Her recommendations and animal handling systems are designed to prevent stress and pain.

In 2015 Grandin received the Meritorious Award of the OIE World Organization for Animal Health, Paris, for her work on developing animal welfare guidelines, one of her many awards and honors. Her books, including *Animals in Translation* and *Animals Make Us Human*, have earned critical acclaim.

Grandin has written, "Raising livestock and poultry is ethical when done right, but the animals must have a life worth living."[20]

occurred. For cattle, vision is the dominant sense for fear memories, though they lack the full spectrum of vision available to most humans because they have only two color cones, as opposed to three in humans. This means that although they are not sensitive to color, they do see the contrast between black and white or dark and light colors.

People who work with cattle can often avoid trouble before it starts. For example, a cow might have had a frightening experience that involved a person wearing a black hat. Subsequently, black hats trigger fright in that animal. The cow might never get over this association, but a handler can avoid problems by wearing a light-colored hat around that animal.

As another example of light-dark sensitivity, animals moving through a barn or other facility will often refuse to walk over a shadow, or step onto a concrete floor from a dirt floor. The high contrast of the shadow or the change in flooring color alarms them; the lead animal stops and lowers her head to see if the pathway is safe. Handlers need to give cattle—particularly the lead animal—time to explore and absorb visual distractions. Once the leader has assured herself that there is nothing to fear, the followers will accept her judgment, and thus a patient handler has avoided a large-scale panic.

Cattle are also affected by what they hear. Some farmers and ranchers frequently play a radio in the barn, which can be helpful because the cattle get used to a variety of sounds that don't cause them harm and have no negative association.

Cattle can handle new experiences, both visual and auditory, as long as they can have time to explore by sniffing, looking, listening, or just coming closer to something of concern.

How Cattle Signal Fear

Whether the cow's response to a new sound or visual goes toward curiosity or goes toward fear is governed by a "switch" in the brain called the nucleus accumbens. As a result of genetics or previous experiences, an individual animal might be prone to either curiosity or fear. So, when the animals are exposed to something new, handlers should be alert for these signs of fear and respond accordingly:

- tail swishing back and forth when there are no flies
- ears "pinned" back (the animal is frightened or angry)
- eye white showing, which indicates intense fear
- defecation (loose manure or soiling of the rear)

At End of Life

If cattle are pressured or frightened on the last day of their lives, this stress affects the color and texture of the meat.[21] Therefore, the meat itself is one indication whether or not the transport and slaughter process went well in terms of humane handling. This connection between the animal's experience and the meat quality provides a practical incentive for handlers to follow animal welfare guidelines carefully.

When cattle are transported to a slaughterhouse, they are moved in a group. Cattle that are docile, and have been well treated, generally walk

Temple and the Cattle Chute

Many years ago, I had just become vice president of a small New England packing plant when we noticed that the beef from one of the cattle we'd killed was noticeably pale, and felt tough even in the package. Something was wrong but I didn't know what. Maybe the problem was with that particular cow or with the farm where it was raised, but because the animal had been slaughtered at our plant, I felt obligated to figure out what was going on so I could pass the information on to the farmer.

The person I turned to for advice was Temple Grandin. Looking back, I realize I was lucky to get an internationally acclaimed livestock expert on the phone. When I described the problematic package of meat, her immediate response was that the animal had experienced stress at some point prior to slaughter. Yes, cattle stress is visible in the meat. The animal could have been frightened while it was being loaded onto the truck or once it arrived at the slaughterhouse. Because the color of the meat was light, it was likely that the stress occurred at the plant. (Dark meat indicates a stressor that could have occurred between twelve and forty-eight hours prior to slaughter; light, tough meat indicates that the stress occurred within twelve hours of slaughter.[22])

Temple has designed livestock handling systems all over the world. Could she come to our plant in Connecticut to check out our animal intake system?

She could not. She was actually scheduled to go to a conference in Connecticut, but her stay there would be busy-busy every minute. I asked how she planned to travel from the airport in New York to the conference in Connecticut, and she said by limo. My response was, "Great! I will be your driver and we can talk on the way." That was fine with Temple.

As an assignment to complete in advance of our meeting, she instructed me to walk with a video camera through the facility's entire intake system, including the chutes, while holding the camera at the level where a cow's eyes would be, so that the film would capture what a cow would experience. I hung up the phone, picked up a camera, and went through our intake chutes exactly as she had instructed.

Days later, I drove Temple and her luggage down the highway while she viewed the film I had made. She spotted the stressor in our system immediately. The spaces between the boards that formed the sides of the cattle chute let in streaks of bright light from the windows. This would be visually alarming to the animals. Temple instructed me to test this by nailing big pieces of cardboard to the outside of the chute to eliminate the streaks of light. I did so early the next morning before the plant opened.

That day there was a marked difference in the behavior of incoming cattle. The plant's staff had grown accustomed to seeing livestock enter the chute with some degree of agitation. But that morning every animal walked in calmly with no hesitation. We corrected the problem, and never again did we see pale, tough meat.

—*Ridge*

calmly, single file, up a ramp into the truck. The process should be unhurried and prodding unnecessary.

Cattle do become nervous in novel situations, so producers can give the cattle some controlled opportunities to be around familiar people, vehicles, sights, and sounds before they leave home on that final trip. In the weeks before slaughter some producers give the cattle the chance to explore the truck that will transport them by feeding them in it. At the facility, some producers keep their cattle calm by walking up the intake chute with them.

Regulations for Slaughter Facilities

There are no federal laws regarding the treatment of livestock on farms or ranches, or regarding the transport of livestock to slaughterhouses. But a number of federal regulations are in place to safeguard the welfare of livestock once they arrive at these facilities. The Humane Methods of Slaughter Act (1978) and associated regulations enforced by USDA's Food Safety and Inspection Service (FSIS) specify proper treatment and humane handling of livestock slaughtered in USDA-inspected slaughter plants.

Federal inspectors are present in slaughter plants continuously. They monitor the stunning process to ensure it is effective in rendering the animal insensible before its throat is slit. In addition, inspectors monitor for other potential animal welfare violations, such as failure to provide water to livestock that are on the premises or dragging a conscious animal.[23]

In 2021, the *Atlantic* published an article by a journalist who had taken a job at a Cargill plant for six months. He began his job there in 2020, but his story opens with an account of an attempted stunning gone awry a full year before. He had not seen the incident but apparently based his description on the

Grandin has written that early in her career she spent a lot of time evaluating whether or not cattle entering a slaughter facility's intake chute sensed that they were going to die. She concluded that they did not; if the intake was well designed and managed, the animals showed no more signs of stress than they would when going into a chute for a vaccination. Her conclusion has been verified by scientists who sampled and measured blood cortisol, the stress hormone. Temple adds that when she gives tours to people through these facilities, they are relieved that the animals remain calm and the death is quick and painless. Malfunctions do occur at times, but these are exceptional.

legally required documentation of it by the federal inspector, which said that the animal had regained consciousness just before it was to be killed, and had to be stunned again before death. The inspector had recorded all the details including the fact that "the timeframe from observing the apparent egregious action to the final euthanizing stun was approximately 2 to 3 minutes." Three days later, FSIS, citing the plant's history of compliance, put the plant on notice for its "failure to prevent inhumane handling and slaughter of livestock," and ordered the plant to create an action plan to ensure that such an incident didn't happen again.[24]

Depending upon the severity of an infraction, an inspector might issue (1) a noncompliance report, which documents the issues, (2) a Notice of Intended Enforcement (NOIE), which permits a plant three days to remedy a problem or face a suspension of operations, or (3) an immediate suspension. When a suspension occurs, a plant might not operate until it has satisfied the agency that it has taken all appropriate steps to correct the problem and prevent it from recurring.

Cattle in a slaughterhouse react to the same kinds of distractions that they might encounter on the farm or ranch. These situations can be eliminated or reduced with attention to the system's design and maintenance, and the training and supervision of facility employees.

In the United States and many other countries, commercial slaughter plants are required to use methods that render the animal insensible before slaughter procedures begin. The options include stunning the animal with a bullet or a bolt or an electrical stunner. The next step is cutting the animal's throat to cause death.

But in the United States and many other countries, religious exemptions to stunning prior to slaughter are allowed, notably the kosher process for Jewish customers and the halal process for Muslim customers. In both these traditions, the animal is restrained and the throat is cut with no stunning. Both religious processes require that the slaughter is overseen by a religious practitioner: kosher by a rabbi and halal by a religious Muslim.

With the exception of these religious exemptions, the animal is restrained and then stunned, often by the use of a captive bolt device (basically a gun that uses a cartridge to send a bolt into the animal's skull), then the unconscious animal is shackled, raised by a winch, and its throat is cut with a sharp knife so that the arteries to the heart are severed, causing quick death by bleeding. The basic process is the same in the small, medium, and largest processing facilities.

Rather than advocate for either stunning or not stunning prior to killing, Grandin raises some issues. She says research is mixed regarding pain without stunning. She feels that the worst problems in terms of pain or stress around slaughter have to do with the method of restraining the animal, and she discusses this issue in her book *Livestock Handling and Transport* and in articles on her website. Regarding the restraining system that holds cattle when they die, she writes, "You have to imagine what it would be like if you were an animal entering it, and you have to respect its every breath, even the last."[25]

Grandin cautions that many people assume that small facilities have better animal welfare practices than large ones, but this is not necessarily true. Even though government employees inspect all slaughter plants, she has observed that the ones that are also audited by a major corporate customer tend to maintain the highest standards of animal welfare.

The Least Harm

A theme of this book is the interconnectedness of everything on Earth. From that perspective, doing the least harm might entail fostering a diversity of plant and animal life, and causing as little suffering as possible. With the intertwined goals of caring for the environment and caring for animals, some people conclude that the most ethical diet for humans is one based on meat and milk from cattle that spend their lives grazing on pasture.[26] As ruminants, cattle are designed to graze; on pasture, their natural behavior can be fully expressed in terms of both diet and movement. A diversity of

Tips We've Learned from Temple Grandin

The most important things to know about handling cattle are what makes them scared or aggressive and what calms them down. Then you can work with their instincts instead of against them.

- All first experiences should be positive for the animal. Cattle never forget a frightening or painful event.
- Animals should be allowed to approach something new—such as a new piece of equipment—voluntarily. Give them time; it is especially important to allow a lead animal time to look, listen, and sniff.
- When driving the herd, as soon as the cattle have started moving in the right direction, back off and relieve your pressure before continuing again. Never apply continuous pressure. Never chase stragglers. They will come back on their own.
- No yelling, whistling, or waving your arms. Animals recognize the "threatening intent" of yelling or whistling. It is even more alarming than the sound of a slamming gate.
- Move animals to a new paddock late in the day when the calves are ready to settle down.
- If hay is temporarily limited, spread it out in a line so that dominant animals don't prevent others from eating.
- To reduce weaning stress, use fence-line weaning or nose-flap weaning so that the nearness of mother and calf can be maintained for a time.
- Keep calves on the farm/ranch until forty-five to sixty days after weaning or vaccination. (Do not wean them by trucking them away!)
- If a cow has experienced a frightening or painful experience, the handler should allow the animal twenty to thirty-five minutes to recover.
- Never let the cattle become feral (wild). Keep up contact with them.

plant material provides the optimal diet for the grazers, and the vegetative cover provides year-round, stable habitat for wildlife, from insects to sentient creatures such as salamanders, foxes, fawns, mice, squirrels, and many more. Grasslands are diverse ecosystems, whereas monocultures of vegetables and grains are not.

Regarding environmental stewardship, regenerative grazing avoids the erosion and climate emissions that are caused by crop production. It also improves soil fertility; protects against droughts and floods; and increases carbon sequestration, resulting in a net climate benefit (see chapter 3). In *Defending Beef*, Nicolette Hahn Niman argues that when calculating the environmental cost of meat production, one should consider the waste and toxicity of producing alternative products, if cattle by-products were not available for use.[27] We have wondered, for example, what might be used for shoes, handbags, and furniture upholstery instead of leather, which is a natural material. Perhaps leather would be replaced by vinyl, a synthetic that is manufactured at great cost to human health and the environment.

In terms of the numbers of animal deaths avoided by various dietary choices, it may be relevant that the death of a single cow can provide two families with protein food for a year, whereas providing equivalent protein from chickens (for example) would entail many more deaths, and providing equivalent protein from vegetables would entail an incalculable number of deaths (as well as suffering) from farm machinery and habitat loss. We are not aware of any diet that is both adequate for human health and also avoids animal death or suffering.

PART 2

Keys to Success with Grass-Fed Beef

CHAPTER 6

Achieving
Wholesale Benefits

A healthy American democracy requires rural communities with vibrant local economies and environmental steward-ship, and farming is at the heart of this vision.
—MARY BERRY and DEBBIE BARKER[1]

We have already seen how raising 100% grass-fed beef has turned some struggling family farms into viable enterprises. Successful family farms benefit their communities. We believe that grass-fed beef production can revive rural economies all around North America and that these revivals will lift up entire regions. But widespread benefits will be realized only when grass-fed beef producers gain access to wholesale buyers—stores, restaurants, and institutions—and much of the region's farmland becomes managed with regenerative practices. This chapter focuses on three essential elements for commercial success with 100% grass-fed beef: (1) cattle that are genetically suited for a grass-fed beef program, (2) a strict husbandry protocol for the care of the cattle, and (3) separate operations for raising young stock and fattening cattle for slaughter.

Genetics and Breeding

Commercial success with grass-fed beef depends in large part on cattle that are genetically suited to be raised and fattened on pasture. While many breeds do well with careful management, it's easiest to build a herd with a breed known to thrive on grass and forage, with no grain, notably one of the following British breeds: Angus, Red Angus, Hereford, Devon, British White, Shorthorn, Belted Galloway, White Park, and Murray Grey.

Within the chosen breed, look for animals that are of moderate height (about forty-eight inches), and are wide and with a deep body; the legs will appear short. We aim for a ratio between depth of body and total height of about 60 percent body-to-height. This is important for two reasons:

- The deep body allows space for a big rumen, which is the fourth part of the ruminant digestive system. Grass-fed cattle get all their energy by processing a great volume of grass and other pasture plants in their rumens. A deep body has space for that to happen.
- Another plus for the deep body is the meat cutout. The goal is to get as much high-value meat as possible from each carcass.

Cattle of one of the recommended breeds with the appropriate body conformation can finish (that is, be fat enough for slaughter) at eighteen to twenty months of age. In a cold climate slaughtering in the fall avoids the expense of feeding the animal stored feed for a second winter and therefore makes economic sense.

Some cattle have a propensity to get fat easily on a grass and forage diet, which is what you want if you are raising grass-fed beef. But throughout the conventional, corn-fed beef industry, from auctions to feedlots, there is a strong prejudice against moderate-sized, easy-fleshing cattle. You can imagine the problem in a feedlot: if one or two animals get fat before all the others in the pen, the condition of those individuals won't jibe with the operator's algorithm for the amount of grain allotted to that pen. The dislike of the easy-fleshing trait almost destroyed certain breeds, including the Devon, which declined to less than two hundred annual registrations in 2001, until the grass-fed beef movement rediscovered their useful characteristics, including easy fleshing.

The decision to sell animals for meat—rather than breeding stock—must be confirmed before other key decisions regarding breeding are made because one needs a clear path for the offspring produced by the proposed breeding. If the offspring were to be used as seed stock to upgrade a herd, the mating pair would be chosen for breed purity and the desired traits of the particular breed. But this chapter is about selling meat to wholesale markets, in which case the producer should take advantage of hybrid vigor by selecting a bull of a different breed from the cows. The resulting calves will be be larger and grow faster as a result of hybrid vigor, but these traits will not be repeated when they are old enough to breed. Therefore, the producer should slaughter and market the crossbred calves and realize a good financial return for the high volume of meat. The producer would then continue to crossbreed for hybrid vigor to get more of those profitable calves.

Two Phases of Production

We noted in chapter 1 that wholesale meat buyers require consistency as well as volume; before we could open accounts for our grass-fed beef with wholesale buyers, we had to nail consistent quality. Our partnering farms already had British breeds, but having the right genetics isn't enough by itself to produce grass-fed beef that reliably delivers a great eating experience.

Eventually we learned that to provide consistent great taste along with volume, we had to address the two stages of grass-fed beef production separately: (1) raising young stock and (2) fattening yearlings. These two stages call for two different farming operations: cow-calf farms to produce young stock, and a smaller number of finishing farms to fatten cattle for slaughter. Perhaps in another part of the country both operations could coexist on the same vast tract of land, but that's not feasible here in the Northeast, where town centers are often only ten miles apart and farms are small. We don't see it happening in other parts of the country either.

We developed an aggregation model that provides both volume and consistent quality. We partner with cow-calf farms all over the Northeast that raise grass-fed young stock according to Big Picture Beef standards and protocol. Then we aggregate yearlings from several of the cow-calf farms to assemble a herd of as many as five to six hundred head on a finishing farm in the same part of the region. Typically, we purchase the young stock from the cow-calf farms when we aggregate them, but other arrangements are possible.

Our partnering finishing farms have larger acreages than the cow-calf farms, and the graziers (generally the farm owners or operators) have demonstrated skills and experience in fattening cattle on grass and forage with no grain. Multiple times per day, the grazier moves the large herd to a paddock of fresh grass as needed, based on an evaluation of the grass and the cattle.

Someone Takes My Advice

Henry came late to farming, after many years in the refrigerated transport business. He began by raising cattle conventionally on his Pennsylvania farm, with the cattle on pasture until they were finished on grain. But eventually he became interested in raising 100% grass-fed beef.

When I arrived at Henry's farm to consult, we hopped in his truck and drove around his paddocks. I confirmed that he had plenty of good grass and could eliminate grain feeding immediately.

But he also wanted to switch breeds, from Red Angus to Devon. He had about a hundred head of Red Angus with good body conformation for thriving on grass with no grain. When I asked why he would want to switch breeds since these cattle were acclimated to the farm and functioning well under his management, he responded, "Well, Devon is the popular breed . . ."

I couldn't argue with that, but I asked to see his Red Angus bulls. He had three bulls and two were not very good for pasture-only production. But one was okay. Henry was in a good position to take advantage of *hybrid vigor*, the biological phenomenon that can result in crossbred calves that outdo their parents by 15 percent in terms of size and rate of growth. I suggested that he keep the best bull and sell the other two. Then I proposed the following breeding strategy to Henry:

- Breed the best 15 to 20 percent of the cow herd to the Red Angus bull to produce the replacement heifers needed annually for keeping the herd at a hundred head; cull old cows and add the purebred heifers back into the herd.

- Obtain a good Devon bull and breed the rest of the Red Angus cow herd to the Devon.
- Then—and this was the key—sell *all* the progeny from those Devon–Red Angus matings for meat. Every single one. Do not put any of the hybrid animals back in the herd to breed. Traits resulting from hybrid vigor do not pass reliably to descendants!

Because of the hybrid vigor generated by a sire of a different breed from the mother cow, I knew the F1 generation (the first-generation offspring from the parents) would be enhanced in conformation, speed of growth, and overall carcass quality. Farmers are always thrilled with these crosses because they are typically are 15 to 20 percent larger than would be expected from a purebred calf. Apparently, the temptation to keep these good-looking heifers in a herd for breeding is an almost irresistible temptation, because most farmers do just that.

Later that same day, another farmer called me saying he had a good Devon bull for sale, and a quick call to Henry had that bull headed to his farm. This was the bull he needed to follow the recipe I had given him for financial success: to realize the benefits of hybrid vigor by selling all the crossbred progeny for meat.

And Henry actually followed the recipe! He stayed in touch, and when he began to harvest the commercial steers and heifers, I asked to see the raw data. His results were excellent. The hybrid vigor advantage had added 5 to 10 percent to his meat yield.

Surely there must have been other farmers who have followed my advice on using hybrid vigor to increase profit, but Henry is the only who followed up with the documentation.

—Ridge

The weather, the time of day, and the point in the grazing season are also factors. Throughout the growing season, the grazier rotates the herd through the farm paddocks, adapting the timing of the moves to all these variables. This skilled finish adds the intramuscular fat to the animals that makes the meat flavorful. Having the cattle finished in a large herd by a skilled grazier ensures consistent great taste.

Addressing the two stages of production separately with the two different types of farm operations was key to opening the wholesale accounts that launched Big Picture Beef, which is a regional business. In our experience, farmers who try to do both in the same operation—to raise young stock and also finish—eventually decide to do one or the other or move one operation to another site. This is the pattern not only in the Northeast but all over the country. Currently there are many cow-calf operations and not enough grass-fed finishing operations.

Basics for Grass-Fed Beef Operations

Some basic principles and practices of 100% grass-fed beef production are the same for both cow-calf and finishing operations; this continuity ensures both stewardship of the land and optimal health for the animals throughout their lives. Because we have a lot to say about a variety of management issues in chapter 7, we're going to focus the rest of this chapter on grazing and one other aspect of production that is key to consistent quality: a husbandry protocol.

Our protocol covers the principles and practices that the two stages of grass-fed beef production have in common, so we will outline the protocol before looking at the differences in the two production scenarios.

Husbandry Protocol

All farmers participating in our grass-fed beef program, both cow-calf producers and finishers, commit to a written protocol that specifies essential aspects of how the animals are to be raised—what must and must not be done. Strict adherence to the protocol is important both for the health of the cattle and also to fulfill the brand claims. For example, "no grain" means no grain ever. Not in the winter, not to coax them into a truck—never. We require producers to sign the protocol.

An affidavit of adherence follows each animal from birth to death so that its history is traceable. We are developing a tracking system that uses an electronic ear tag to track an animal's whereabouts during its entire lifetime. It is the law under the European Union to determine the location of animals at all times, and there are several digital programs that track the animal and allow a producer to print a passport at harvest to travel with the meat to butcher shops or stores.

The protocol is important for our wholesale customers' public relations, too; their customers might be coming to their store or restaurant because of a particular health or environmental practice that is the basis of their patronage (for example, no antibiotics). Therefore, adherence to the protocol's provisions is not negotiable. The full document is available on our Big Picture Beef website (bigpicturebeef.com); what follows is a summary:

- pasture for regenerative grazing: grasses, legumes, forbs, and planted cover crops from fields that are non-GMO and free of biocides for three years
- stored feeds: haylage and baleage (chopped grass and other pasture forage that is stored in wrapped bales for fermentation)
- calves on mother's milk, preferably, for ten to eleven months (a minimum of six months)
- free-choice minerals (optionally)
- no grains, industrial by-products, or animal or fish by-products
- no corn silage
- no antibiotics or ionophores
- no growth hormones or implants
- identification of all calves from birth
- a signed affidavit of adherence to this protocol accompanying any animal that is transferred from one farm to another, or to a finishing farm or slaughterhouse, required

Offering minerals is optional in the protocol, but we want to note that minerals impact all bovine hormonal functions, particularly reproduction and growth. Lack of specific minerals can lead to a host of typical herd health problems, such as pink eye, lice, flies, foot rot, retained placenta, and poor calves.

While long roots can access soil minerals that pasture plants can use with the help of soil microbes (perennial prairie plants, for example, grow ten feet deep or more), recreating the conditions of a prairie doesn't happen overnight. Offering free-choice minerals can offer some assurance that the cattle have all the minerals they need for optimal health, especially at times of stress. Using a cafeteria-style mineral dispenser may seem expensive but could preempt herd health problems before they develop.

Grazing Specifics for Cow-Calf Farms

Although both our cow-calf producers and finishers use regenerative grazing practices (adaptive multi-paddock grazing), cow-calf producers do not need to move their cattle nearly as often as finishers. The nutritional needs of the cow-calf pairs are different from the nutritional needs of the cattle that are being fattened for slaughter. This section addresses feeding the cattle on the cow-calf farms throughout the year. These farms must manage grazing for the five classes of cattle: cows, calves, bulls, steers, and heifers. We describe grazing planning for all five during the grazing season. We also address winter feeding for cold climates that necessitate stored feed when the grass has stopped growing.

Grazing by Cattle Class

For the cattle on a cow-calf farm, there are two things to consider regarding nutrition: the seasonal grass cycle and the class of cattle.

In the Northeast there is tremendous growth of grass beginning in April and going through June. Then as the heat increases and the rain decreases, there is a slump in growth until the end of summer. Then grass begins to grow again and there is a flush of new growth until early October. The farmer needs to match the nutritional needs of each class to the pasture resource throughout the growing season. To understand the various needs of the cattle in a cow-calf operation, it is useful to think of the five classes of cattle in three groupings: (1) cow-calf pairs, (2) weaned calves (both heifers and steers), and (3) bulls.

COWS-CALF PAIRS

Cow-calf pairs take far less management than finishing steers and heifers; their needs fit well with a part-time farming schedule. The cows are bred to

Prepotency: Apples Fall Close to the Tree

Here is an example of the effect of using a *prepotent* bull—that is, one that consistently passes on his traits to his offspring.

The late New Zealand breeder Ken McDowall sold semen from a prepotent Devon bull, Rotokawa #688, to numerous ranchers in the Northeast and Western United States and Western Canada. Six thousand cows were bred with semen from that handsome bull. When a big calf crop was on the ground, Ken and I visited some of the Canadian ranches that had used Rotokawa #688 semen.

One twenty thousand–acre ranch we visited had been using quite a few bulls over the previous fifteen years and as a result the cow population that we saw spread across the fields was of the Heinz 57 variety. But as Ken and I drove along the farm roads looking at the vast herd, we found that even at a distance we could easily pick out the deep-bodied calves sired by #688, and then verify our accuracy when we got close enough to read the ear tags. We were correct with a consistency that was stunning even to Ken. These visually outstanding calves were a dramatic example of the impact of a prepotent bull on his offspring.

The only way to determine that a bull has the prepotency trait is to evaluate his calf crop to see if every single calf resembles the sire; this is how Ken McDowall determined that #688 had what it takes.

—Ridge

produce a calf every year. Because the gestation is nine months, we breed in August through September so that calves are arriving in late April through May, when the grass has started to grow here in the Northeast. This provides a nutritional boost for birth and milk production. The calf should stay with the cow until ten or eleven months, which is at least a month short of when the cow will produce the next calf.

Because the cow is a fully grown bovine, she has had both the protein and the energy from her pasture diet to reach optimal height and weight. (In this context, it's helpful to understand that the bottom of a pasture plant is the protein portion and the top provides energy.) Therefore, the cow requires only a maintenance ration of grass and forage, enough to support herself and to produce milk for her calf. In contrast to its mother, the calf needs protein for long bone growth, and it will get this protein both from the mother's milk and from grass eaten near the base of the growing plant. The important thing is that the cow-calf pair have enough to eat. If the grass is shorter than four or five inches, the pairs should either have a larger area or be moved more frequently to a new paddock.

If paddocks are sized with enough grass, cows and calves can be moved to a new paddock every three days. They should not remain longer because the grass will begin to regrow and should not be regrazed.

Cows that have already weaned their calves will also do well in "landscaping mode"; that is, they can get all the nutrition they need from cleaning up brushy areas that are not suitable for calves, yearlings, or bulls, all of which need to be on a higher plane of nutrition.

WEANED CALVES

After calves are weaned and their long bone growth is complete (they reach their full frame height close to a year of age), they are like adolescents that no longer need their mother's company and are ready to join a larger group of age peers. This age group needs a ration with a higher proportion of energy in order to put on flesh and begin to store fat. In other words, they now need access to the tops of the plants. As a transition period in preparation for them joining a larger herd of finishing cattle, some graziers use a leader-follower system by which the yearlings move together through a paddock first, eating the high-energy plant tops before moving on to the next paddock; then the cow-calf pairs move into the paddock that the yearlings have vacated. (Think of the yearlings as skimming the cream, and the cow-calf pairs following to get the milk.)

BULLS

Bulls need enough energy from grass to maintain their body condition and to breed cows. Our recommendation is to allow the bulls to grow close to their mature size and weight (two years and three months) before putting them to work breeding. Waiting for them to mature ensures the longevity of the breeding stock. If the grass is eaten down to less than four or five inches, the bulls should either have a larger area or be moved more frequently, as described for the cow-calf pairs.

Winter Feeding

In chapter 7 we will discuss *stockpiling grass* (leaving grass and forage uncut and ungrazed in the field) but will touch here on cut grass stored for winter as dry hay or baleage (fermented grass in a plastic-wrapped bale).

Feeding stored forages is the most economical way of accomplishing two things at once: nourishing the cattle and applying nutrients to the land with little mechanical assistance. One option is bale grazing: depositing either dry bales or baleage in areas around the pasture that are well drained and can use the additional impact of nutrients. Then throughout the nongrowing season, the bales can be opened with a jackknife rather than using a tractor daily to move feed to cattle. Yes, there will be some material that is not eaten, but this is not waste; the nutrients get spread across the farm rather than collecting in one area where they would eventually need to be distributed with a tractor and manure spreader. (The only waste problem with baleage is that recycling the plastic used to wrap the bales is still technically challenging. Though attempts are being made to improve the effectiveness of recycling agricultural plastic, even the best programs have not perfected the process.)

Another option for farmers is using a bale unroller to unroll dry, unwrapped bales in areas that need an additional boost of nutrients. The unroller can be pulled by a four-wheeler or a pickup truck. The bale might be spread out over two hundred feet depending on its size.

Cows do very well on dry hay; some people prefer it for cows because they can get too fat on baleage. Baleage is the first choice for heifers, steers, or bulls.

Grazing Specifics for Finishing Farms

As we've explained, coordinating groups of cattle that have various needs for nutrition can be a challenge; therefore, the planning and implementation of finishing becomes much simpler with only finishing cattle rotating through the available paddocks.

Aggregation: A Once and Future Model?

In the seventeenth century, Europeans invaded the Northeast and settled in the shared homeland of its Indigenous populations, the Iroquoian and Algonquian peoples, replacing their villages, small fields, and forests with English house lots, villages, and modes of agriculture. The Europeans reshaped the landscape by clearing land and fencing farm fields. What follows here is our reflection on one aspect of the agricultural system that they developed as colonists—aggregation of farm products for marketing—and a brief outline of what happened to that marketing concept in the next two hundred years.

By the early 1800s, New England farm families of European descent typically had a one hundred–acre farm and four or five cows. Each family milked their cows, used their castrated males as oxen, and eventually ate their animals.

As this was prior to refrigeration, during the growing and grazing season, farm women turned the abundance of milk into products with a longer shelf life: butter and cheese, which was taken to the general store and sold or traded. The storekeeper was an aggregator, storing cheese, for example, in a barrel until enough was accumulated to put in an ox cart and take to a big city market like Boston's. Livestock was also aggregated from many small farms and driven in herds or flocks to the larger city markets.

As the decades passed, Northeast farms became larger and more specialized, and train tracks were laid throughout the landscape. Milk and dairy products were accumulated from many small farms and transported by train to city markets. This aggregation and distribution system was the norm well into the twentieth century.

When the two of us moved into the central Massachusetts town of Hardwick in the 1970s, thirteen dairy farms were still milking cows. By that time, a tank truck that went from farm to farm to get milk from those thirteen dairy barns had replaced train transport, but we remember local elders reminiscing about their youth in the 1920s and '30s, when they delivered big milk cans to the railroad siding, where they were picked up by "the milk train."

Advancing rapidly to the present—and skipping over the federal farm policies of the '70s and '80s that shaped the present rural landscape—there are no more dairies in that town. As one drives through the rural Northeast, there are many vestiges of the not-so-distant agrarian past: fine public buildings and houses that were built with proceeds of the robust agricultural economy here in the nineteenth and early twentieth century. Much of the land that was formerly grazed by cattle and sheep is now abandoned for agricultural use. But farther west, the United States now has feedlots with thirty thousand to forty thousand beef cattle, and mega-dairies with as many as ten thousand dairy cattle on a single farm.

Assuming that aggregation of livestock farm products doesn't have to lead to the unsustainable and unjust imbalances that characterize modern livestock production, we have a different vision of rural prosperity. We advocate for a system of producing grass-fed beef cattle that includes numerous small cow-calf operations raising yearlings that are then aggregated on a smaller number of finishing farms in the same region. With this system, widespread regenerative production could restore degraded farmland to former health; revive rural economies; create new, skilled, agrarian jobs (such as finishing cattle on pasture); and provide locally produced grass-fed beef for each region's urban populations.

—*Ridge and Lynne*

A skilled grazier can evaluate yearlings at the beginning of their second grazing season to determine the probability of their gaining enough weight to be slaughtered before the following winter. Those yearlings that are at least 750 pounds going into the season will likely be heavy enough, but even lighter animals might be ready if they have a good rate of gain. All will spend the second grazing season in a large herd of yearlings that will be moved frequently (multiple times per day) to a new paddock throughout the spring, summer, and fall. Steers will be heavier than heifers at the same age, but heifers will finish (get fat) at an earlier age because they fatten more quickly, albeit at a lower total weight. Because experienced graziers anticipate variability in readiness for slaughter, they set up a staggered schedule for slaughtering.

Finishing: The Steps

Finishing (fattening) cattle for market requires more of a time commitment and intense effort than raising young cattle; it is a full-time job. Getting cattle fat on grass and forage alone (no grain) requires moving them frequently, perhaps several times a day, to a fresh bite.

Optimal care of cattle and pasture calls for short periods of grazing and much longer periods of regrowth and rejuvenation for each paddock. In the simplest terms, the following three steps describe the way graziers fatten cattle without grain:

1. A knowledgeable grazier draws a grazing plan that shows the available grassland divided into the number of temporary paddocks likely needed for a multi-paddock rotation throughout the season.
2. The grazier moves the cattle to a new paddock when less than the top half of the plants have been removed, with each new grazing area created by moving the flexible fencing. A *tumble wheel* is a simple device that makes it possible for just one person to set up a new paddock and move the herd into it. (The

The key to increasing the rate of gain is to present a lot of energy (tops of plants) on a constant basis. The young stock know that energy is what they need and will quickly eat the tops of the grass in a new paddock. The management challenge is to have a new bite of grass in front of them continuously. Energy levels in the grass are higher later in the day after the sun and photosynthesis have been working, so the grazier can take advantage of this fact and move them into a new paddock in the afternoon.

Note that some young stock will not be finished at the end of their second grazing season and will need to go through a second winter, finishing on stored feeds at twenty-two or twenty-four months, instead of eighteen to twenty months. Baleage is an easy way to feed them through the winter.

> animals must also be fenced out of the paddock that they have just grazed to prevent their return.)
>
> 3. Eventually the cattle cycle back to the original paddock but *only* after the plants (and roots) have fully regrown. The regrowth period is variable depending on how fast the plants are growing, and their stage of regrowth. In a dry climate, a paddock won't be ready for cattle again until the next year, whereas in an area of abundant rainfall, such as the Northeast, a paddock might be grazed again in a month. The grazing rotation tends to slow down during the course of the season.
>
> The word *adaptive* in the phrase "adaptive multi-paddock grazing" is well chosen. The three steps of finishing as we've described do not convey the knowledge and skill required to make an effective plan and to adjust it to the needs and condition of the paddocks and the cattle on any given day. Moving the cattle is always a judgment call based on the status of the pasture plants and the animals.
>
> For details on making and adapting a grazing plan, we highly recommend Sarah Flack's *The Art and Science of Grazing*.

The components of 100% grass-fed beef production that we've outlined in this chapter provide the foundation for success:

- choosing a breed that will flourish on grass and forage without grain
- following a strict protocol for care of the animals and management of their pastures
- keeping the two phases of production—cow-calf and finishing—separate, which usually means making a choice between the two for any one farm or ranching operation
- making a regenerative grazing plan and then adapting it to changing conditions

If all these principles are followed carefully, the result should be consistent, high-quality beef. Aggregating the grass-fed yearlings from a number of cow-calf farms for finishing can provide volume. Thus, the wholesale buyers' requirement for both volume and consistency can be addressed, so that producers can access dependable wholesale markets in their region: stores, restaurants, and institutions.

Wholesale accounts for regeneratively produced grass-fed beef also ensure that multiple benefits accrue for communities throughout a region. Money spent on regional beef will stay in the region. Family farming can provide a viable income. Farmland will be rejuvenated. Streams will no longer carry runoff from biocides and chemical fertilizer. Both rural and urban areas can enjoy the health advantages of 100% grass-fed beef. Landscapes will reflect flourishing rural economies once again.

CHAPTER 7

Turning a Profit

Obstacles are those frightful things you see
when you take your eyes off your goals.

—ANONYMOUS

T his chapter is about dollars and cents. Every businessperson wants to make a profit, but for people who raise livestock, profit margins are often very small; a bad year can jeopardize the whole enterprise. Furthermore, people who have no inherited assets or access to capital may feel that raising beef cattle is entirely out of reach. But raising grass-fed beef offers some new opportunities for livestock producers to succeed.

Because regenerative grazing is based on working with natural systems (and not against them), it is possible to start raising grass-fed beef with minimal capital. A farmer or rancher can achieve net profitability by fostering the biological activity that we described in chapter 3; these practices will improve the soil's health and keep the costs of production low. Profit will come from the health and productivity of the fields and the cattle. Producers will know they are on track for profitability when the organic matter in their soil increases, their paddocks support more cattle, and those cattle are healthy. Improved soil fertility and structure will nourish the pasture plants and the cattle, and improve the bottom line.

Minimal Capital Investment

If you can afford to purchase land, build a barn, and buy a tractor and haying equipment—and you want to own these things—then go ahead and buy them. But consider this: You don't need to own all of them—or any of them—to raise grass-fed beef.

Renting Pasture

Of course, you'll need pasture for grazing. But that doesn't mean you have to buy it. Depending on your community, pasture land might be available to rent with a low-cost or free lease.

CONSERVATION LAND

A great deal of acreage is owned by conservation groups, states, and other public entities. Sometimes these land stewards find that once grassland has been saved from development, the effort to keep it from growing up to brush is a maintenance headache; if you leave land alone, brush becomes trees. To keep farm fields open, some commonly employed methods are burning (which costs thousands of dollars per acre) and spraying with poisons—most likely glyphosate—with risk to health and the environment. Some conservationists simply hire a Bush Hog (rotary mower), at considerable expense, to cut down the grass, and then simply leave it lying in the field.

But one farmland stewardship option is win-win: regenerative grazing.

Getting a long-term lease to graze conservation land requires some conversation with decision makers about their expectations for how the property will look. They need to understand that healthy, diverse grassland might include a diversity of "weeds." And the tall grass that is perfect for grazing might look to some like hay gone to waste. Be proactive: provide the community with information about the importance of letting the grass grow tall and the roots grow deep so that soil life can perform its functions.

The National Audubon Society has initiated a program called Conservation Ranching to certify land that is used for grazing cattle in a way that protects birds and other wildlife. This program may help to inform the public about the compatibility of grazing and wildlife conservation.

LEASING FROM HOMEOWNERS

Many homeowners love a pastoral landscape and would like to keep their fields from filling in with brush. But like the conservation groups, they, too, might have some unrealistic expectations of how a pasture looks at different times in the growing season and before and after grazing; so when you make an agreement, let the landowners know that the pasture will be managed to ensure its long-term health and good nutrition for the cattle—and what that management will look like.

DAIRY LAND

Because the price of milk remains low but the costs of production have risen, many dairy farmers are looking for other ways to earn income from their land; some are willing to rent pastures for grazing.

Single Wire Fences

With grass-fed beef, the need for permanent fencing is minimal. You do need a sturdy perimeter fence to keep the animals off the road and off the neighbor's farm, but interior fencing can be temporary: a single electrified wire that can be easily moved, perhaps with the aid of a device like a tumble wheel if the terrain allows. Remember, with regenerative grazing the fence is more to designate a new area for grazing than to confine animals; if cattle always have grass in front of them or are comfortably full, there is no pressure on the fence. Having flexible fencing allows you to apply animal impact differently at different times. For example, you can create a new lane for the cattle to walk to their water supply if the route they have been using gets muddy.

No Barn

Grass-fed beef cattle have no need of a barn and will stay outside if given a choice. One problem with a barn is that it tends to accumulate manure and urine, concentrating the nutrients that are produced in a space with an impervious floor, which leads to runoff and groundwater pollution. Beef cattle are healthier outside where the sun shines. Do provide a sheltered spot where they can get out of the wind, be it in a valley or by the woods, a stone wall, or a building. Given good grass and all the water they need (a backup supply of water is especially important for a large finishing herd), the cattle will flourish, even in a northern climate.

Of course, not all breeds of cattle are this robust. For example, most conventional dairy cattle have been bred to a large size, but they lack the capacity to maintain their body condition while they produce milk, unless they are in some kind of shelter. In contrast, the recommended beef breeds are rugged animals, much like their wild ancestors. Their bodies are very different from human

THE LEARNING CURVE

Bullhorn Diplomacy

In the early years of the millennium, I was hired as a consultant to the Argentine government on grass-fed beef production. This was a World Bank initiative, so instead of sending me to the Pampas grasslands where there was already a large, successful grass-fed beef industry, they sent me to the northern Chaco region, which was struggling economically. The trip entailed several formal presentations, but I requested that we also visit a farm for a field day, where I could demonstrate theoretical points in a real-world scenario.

As soon as we arrived at the ranch, we went to view the cattle in the grassland. I had an interpreter who stood near me and translated through a bullhorn to the gathered group of 150 farmers. We stood near the fence and watched the gauchos on horseback bring the cattle up closer to the fence.

I was shocked when I saw these animals. Even though this was the growing season in Argentina, the cattle were not in good condition; I could see all their ribs. They were Brafords (Brahma crossed with Hereford) and Brangus (Brahma crossed with Angus), and the Brahma influence had made them much too tall for the forage available there; that poor-doing grass would never fill out those gigantic frames. Glad for the time delay afforded by the translation process, I did my best to say something about the cattle that would be useful to the group without embarrassing the rancher.

Later we moved on to *linear measure* some of the cattle, an exercise that highlights both strong points and deficits of an animal's body conformation in a way that allows us to predict whether the animal will thrive on a pasture-only diet. This activity allowed me to base comments about the rancher's herd on an objective standard and talk to the group about what type of cattle would best suit the grass in this location.

Later in a private moment, I questioned the rancher about his choice of bovine genetics. He responded, "Well, I got the semen from a popular bull stud in America." Knowing that the Brahma are used because of their heat tolerance but were also known for tough meat, I asked him where he marketed the beef. He responded that nobody in Argentina would buy it or eat it, being used to pretty good grass-fed beef from the Pampas region, so he sold it all to Brazil.

So, this was a case where the wrong genetics and breeding led to poor production and a low price in the marketplace. Clearly this gentleman wanted a cattle ranch and was willing to subsidize his endeavor. Cattle that are ill suited for a diet of grass and forage would be of little help in developing the health and fertility of his land to the point where he could turn a profit.

—*Ridge*

bodies. Whereas a person without proper gear might not survive a stormy winter night in New England, one of our grass-fed cows, with her heat-generating rumen, insulating blanket of fat, and warm hair coat that retains body heat, is perfectly content to lie down in a snowy field on a ten-degree night. She gets up in the morning, shakes off the snow, and carries on with her day.

Some producers do provide a covered feed bunk in winter, perhaps not so much to cover the cattle as to keep hay from getting soaked. Such a structure does not need to be a fully enclosed building.

Tractor? Haymaking Equipment?

It's helpful to enumerate the specific tasks throughout the year that call for a tractor and consider whether those jobs justify the considerable expense of buying, maintaining, and running one. If the need is minimal, it may be more economical to rent a neighbor's tractor or hire the neighbor to spread bales around the farm in the fall before winter weather makes it challenging to move bales.

Likewise, do the math on your costs for hay. Whether it is homemade, made on your land by someone else, or purchased, this expense might be the difference between a profitable operation or not. Where we live it costs about $50 a bale to buy hay and $40 a bale to make hay if you include fuel, equipment, depreciation, and labor.

When deciding on the best use of the fields available to you, consider this information. A study of ten livestock operations in four states in the Southeastern United States compared adaptive multi-paddock grazing to conventional grazing: the regenerative approach increased the biomass generated per acre threefold.[1] The regeneratively managed acres were able to carry more livestock, and even with the higher stocking rate, these operations had better water infiltration, more plant diversity, and higher soil carbon levels than neighboring operations that were managed convention- ally with continuous grazing. More livestock means more income. Also, the water infiltration, diversity, and carbon levels indicate land that is functioning better and so needs fewer purchased inputs to be profitable.

What if you sold your haymaking equipment, grazed cattle regeneratively on the land that was previously hayed, reaped the benefits of increased forage in your pastures—including raising more animals on the same land base—and focused on extending the grazing season? You wouldn't need to buy much hay.

The next section explores techniques for making your pastures more profitable.

Pasture Management

Although a number of things are needed to ensure success with grass-fed beef (choice of cattle breed, a well-informed grazing plan, and so forth), perhaps none are more important than excellent pasture management. Two proven techniques for getting more from your pastures are planting cover crops for extending the grazing season at both ends, and planning ahead for winter grazing. (Yes, even if it snows where you live.)

Cover Crops

Planting cover crops is one of the most valuable ways to feed grass-fed beef cattle well and inexpensively. Cover crops can augment your perennial graz- ing three ways: (1) by emerging before the perennial sward, thereby starting the grazing season earlier, (2) by enhancing the usual "summer slump" of perennials suffering from heat and lack of rain, and (3) by extending high-energy grazing into the typical nongrowing season. (By "high-energy," we are referring to the energy at the tops of the plants, not the energy of the cattle!)

A "cocktail" cover crop is a mix of eight or more plant species that can work well on a finishing farm, especially to convert idle, misused, or "farmed out" land back into vibrant perennial grassland. This approach can provide significant production and income while the land transitions into a diverse polyculture of grasses and forbs.

We advise sending a soil sample to a lab that tests for biological activity in the soil, rather than for minerals, which is the more common test. Over time the use of cover crops can reduce or eliminate the need for fertilizers and herbicides—along with their expense. Research scientist Rick Haney, the creator of the Haney soil health test, said he originally developed the test to sustain yields while cutting back on fertilizer. Most of the nutrients that ulti-mately end up in a crop are influenced by soil microbes. In some cases, the microbes either temporarily tie up or release nutrients based on their own metabolic requirements and the health of their ecosystem. The Haney test results indicate the amount of food that is readily available to soil microbes. It measures root exudates and decomposed organic material.[2]

The goal of Haney's approach is to help farmers track their success in building soil health, but the costs savings in reduced need for fertilizer is another incentive to use the test and plant a cover crop. Of course, there are costs associated with planting it—seed and machinery and a no-till drill to seed the ground. A cost-benefit analysis is necessary to determine which annuals to plant and the timing of the planting to synchronize with the perennials already in your pastures.

Some species (like chicory) are particularly high in energy, and this crop also has a deep tap root that helps break the "plow sole" (a compacted layer of soil about 8 inches below the soil surface, created by plowing with a mold-board plow) that exists in many formerly cropped fields. Some choices, such as rye, start growing early in the spring (if fall planted) and create a highly nutritious "early bite" to get the grazing started earlier while waiting for the perennials to gain some height to produce more energy. (Wait for the third leaf.) Remember, the finisher's job is to offer the cattle enough volume of biomass as well as quality (energy).

Another significant factor to consider is the improved weight gain per day from the cover crop. A high rate of gain is key to finishing cattle in the early months of the nongrowing season before the second winter. The difference between 2.25 pounds of gain a day and 1.8 pounds of gain represents significantly more costs later, because cattle gaining less weight (1.5 to 1.8 pounds per day) will take twenty-four months or more to be finished, which means a second winter of using stored feed. Therefore, the addition of a cover crop to extend high-energy grazing into December and January can significantly decrease costs of production.

Resources for Measurement, Evaluation

There's merit to the saying that if something is important to you then measure it. We have mentioned the value of soil tests like the Haney test, which evaluates what quantity of soil nutrients are available to soil microbes. While a comprehensive list of resources is beyond our scope here, we will mention a few measurements that could influence your planning and eventually your profitability:

A plant tissue or sap analysis tells you what is in the forage and what is missing. For this you need to send a sample into a lab to get a report. Deficits in the plant may indicate treatment with a foliar spray to stimulate the plants. The healthier plants are then better equipped to carry out more photosynthesis and thereby stimulate the soil biology.

A Brix meter (refractometer) measures sugar in plant leaves. This is an inexpensive tool that you can carry in your pocket. Grass with a high Brix level leads to good nutrition, faster-growing cattle, and more fat deposits that give the meat flavor.[3] An important caveat: using this tool to make management decisions requires a great deal of experience with it on a given piece of land. Brix levels fluctuate through each twenty-four-hour cycle, usually peaking mid- to late day because of accumulated compounds created by photosynthesis. In healthy plants with the proper mineral balance for good transport of these compounds, Brix levels often drop 30 percent or more in the leaves from evening until morning, as sugars are used or stored.[4] Learning to interpret the readings is a useful skill to acquire.

Cocktails for Cows

My first experience with cocktail cover crops was a dozen years ago, consulting for a farm in North Carolina that had a herd of beef cattle of mixed British breeds. About one hundred of the farm's three thousand acres was a burned-out tobacco field. The farm owner was willing to try regenerative grazing if I could make it profitable on that piece of land.

Someone suggested that I call regenerative farmer Gabe Brown in North Dakota for advice, so I placed the call and happened to catch him driving with soil expert Ray Archuleta in the passenger seat. After hearing about my challenge with the unpromising tobacco field, they suggested that I take a soil sample and send it for analysis to soil scientist Rick Haney, the creator of the now-famous Haney test.

So I sent my sample to Haney's lab and the results came back with plant recommendations for a cocktail cover crop mix formulated specifically for that acreage. We bought the seed, mixed it, and planted it with a no-till drill. When it was grown, we began to graze it with the farm's 150 finishing steers and heifers. We sectioned the pasture with electric fence and let the cattle into a new section a couple times a day. The cover crop was over their heads.

The results were stunning. The feeders gained over three pounds per day (equivalent to feedlot gains on corn) on burnt-out land. We also observed the increased biological activity of the soil, in terms of earthworms and other visible soil life.

Looking back, I realize how lucky I was that day to connect with Brown, Archuleta, and Haney—all celebrated now for their outstanding contributions to regenerative agriculture—who led me to do the right thing for that worn-out tobacco field.

—Ridge

Grazing in January

I had been interviewed for *TIME* magazine on the importance of grazing ruminants to combat climate change.[5] To illustrate the story, *TIME* was to send someone shortly after Christmas to get a photo of me with cattle. The photographer happened to come on one of the coldest days of the year.

The previous summer—long before I knew I would be featured in *TIME*—as an experiment I had kept the cattle out of one of the paddocks after July in order to let the grass grow long for winter grazing. Now that paddock was covered with about eight inches of snow. I thought, wouldn't it be great to have a photo in *TIME* that showed the herd grazing in January?

As soon as the photographer arrived, I let the cattle into the paddock with that grass I'd stockpiled the summer before. As if on cue, the cattle nosed right into the snow to get at that grass, dry and brown as it was. They grazed during the entire photo

A **soil food web assay** is done by using a high-powered microscope to count the soil food web organisms. Dr. Elaine Ingham has popularized this method, and you can send samples into one of her labs to get a report. Her method to boost biological activity is making specific compost and compost teas suited to the soil.

Winter Grazing (Stockpiling)

A strategy for extending the grazing season for cows and calves is stockpiling grass for winter, by letting an area remain uncut and rooted in the ground. It will be dry and brown and perhaps covered with snow, but cow-calf pairs will love it, and their manure will have the appearance of pudding, indicating good rumen functioning.

shoot, rarely lifting their heads. By the time the photographer was satisfied with his shots and had left, hours had passed and I was stiff with cold. The cattle were still grazing.

I went over and opened the gate so the animals could retreat into the adjacent paddock where I had put a big round bale of hay. But they had no interest in the hay; they kept grazing that stockpiled grass through the snow. It was the same color as the hay, only still rooted in the ground.

The *TIME* article came out in January 2010, but the day that the cattle had preferred the stockpiled grass over hay stuck in my mind, and a couple of years later moved me to apply for a grant from SARE (Sustainable Agriculture Research and Education) to conduct a trial of winter grazing with my cattle, which I did in 2013. By having samples of the stockpiled grass analyzed, I discovered that winter grazing provides excellent nutrition for cows and calves. I already knew that cattle will nose through the snow to get it.

—Ridge

We undertook a trial of this method by selecting twenty acres of pasture, which we stopped grazing in July. Because we had received a Sustainable Agriculture Research and Education (SARE) grant to do this study, we followed a protocol to test the nutritive value of stockpile. To create a baseline, we took forage samples in July and sent them to a lab (Dairy One) for forage evaluation of the things that are typically measured to determine the quality of feedstuffs, such as the total digestible nutrients and relative feed value.

We brought the cattle back into the stockpiled area to begin grazing in December. We used a single wire electric fence to portion off just enough for them to eat each day through December, January, and February. We also took photos and forage samples for each of those months at six designated locations, sent the samples to Dairy One, and plotted the nutrition values on a graph. Through the winter months the values we measured declined slightly, as we expected, but when we compared them to the values of a bale of hay from UMass and the values of baleage from a local farmer, we were surprised and pleased to see that the nutritive values of the grass in our stockpile were higher than those of the hay and baleage.

By spending some time with a calculator you can determine the cost per day for whatever stored feed you use presently. Although there are some minimal costs involved in winter grazing, the costs savings for every day you can feed stockpile instead of hay or baleage is considerable and could be the difference between breaking even or making a profit.

Cattle Management

Of course, profitability with grass-fed beef depends very much on the details: how you care for your animals, how you manage their breeding, and how you address their nutrition needs throughout their lives. What follows are some management issues that affect the bottom line but are often ignored.

Raising Moderate-Sized Animals

We have already recommended moderate-sized British breeds for grass-fed beef production, because of their early maturity and ability to fatten on grass and forage. But there's another way in which these moderate-sized animals bring you more pounds, which means more income.

Let's say you have seventy-five acres of pasture, and on average you produce three tons of forage per acre. Same is true for your neighbor. Cows need approximately 3 percent of their body weight per day for nutrition to maintain their body condition and raise a calf. So, whereas your neighbor's 1,600-pound Simmental cow needs 48 pounds of dry matter per day, your 1,200-pound Devon cow needs about 36 pounds of dry matter a day. Since your tonnage is three tons (6,000 pounds) per acre times 75 acres (450,000 pounds) you can support 34 of your moderate-sized cows on your spread, whereas your neighbor can support 25 large cows. Each of those cows is going to produce a calf every year. Your cows will produce 34 calves, your neighbors will produce only 25. Because you have a moderate-sized breed, your herd will yield more salable pounds than your neighbor with the larger cattle on the same size land base.

Letting Cattle Mature

The guiding principle of working with nature whenever possible applies to breeding cattle and raising calves. A program that accommodates the animal's age, size, and instinctive behaviors is not only humane but also improves profitability.

BULLS

Bulls need a fair amount of energy from forage to maintain their body condition and also breed cows. As noted in chapter 6, we recommend allowing the bulls to grow to close to their mature size and weight (two years and three months) before putting them to work breeding. This ensures longevity of the breeding stock.

HEIFERS

Farmers/ranchers expect their cows to produce a cow every year; in fact, when cows fail to conceive the year after having their first calf, they are typically culled from the herd—an investment loss. We believe the frequent failure of heifers to "breed back" (typically 15 percent of them) is because heifers are often bred when they are too young to begin their reproductive lives. A heifer should be three years old before she has her first calf, which means breeding her at two years, three months. Then you can expect her to have a calf every year. If you have a British breed and let your females fully develop before carrying a calf, your cows may have nine calves in their lifetimes, paying back handsomely your investment in them.

CALVES

In conventional production spring calves are often weaned at five or six months, going into winter, because some mother cows cannot support their own condition plus feed a calf in the nongrowing season. But weaning this early is stressful for both mother and calf; the literature is full of strategies to minimize the stress. And there is a physiological issue; the calves are typically moved to stored feeds upon weaning and experience a drop in growth as their gut flora switch from grass and mother's milk to hay or baleage.

These problems can be eliminated simply by leaving the calf with the mother. Regarding the mother's condition, a rugged British breed cow with the conformation that correlates to success on a grass and forage program will have

Opportunity for Dairy Farmers

All across the country, and certainly in the Northeast, dairy farmers and their families are in a desperate situation. In 2020, the USDA reported that the number of licensed dairy farms had declined approximately 30 percent in the previous decade, a long-term trend caused by low prices and an oversupply of milk.[6]

Given the grim challenges faced by dairy farmers nationwide, many are considering other options for their livelihood: (1) converting from a conventional dairy farm to a grass-fed dairy farm, (2) raising and selling heifers to other dairies, (3) making and selling hay from the farm fields, or (4) switching from producing milk to producing 100% grass-fed beef.

The first three options are possible but problematic. Dairy farmers face the challenge of finding dairy cattle with the deep body that indicates the capacity to process large amounts of grass and forage. Selling heifers to dairies only works if the buyers manage to stay in business despite low milk prices. Making hay is a low-margin business, and continuously removing hay without returning organic matter to the soil can feel more like mining the land than farming.

no problem nursing her calf for ten or eleven months. By doing this you can also expect 15 percent more growth in the calf over the winter, which means the calf will fatten by the end of the year, which we have noted is a financial boon.

Rewards of a Separate Finishing Operation

There is a widely held but incorrect idea that grass-fed calves take twenty-four to thirty months to get fat; conventional producers and the industries behind them use this misconception to argue that grass-fed beef production is uneconomical. (They also assert, incorrectly that these animals are adding to the problem of methane accumulation for those extra months.)

We believe the fourth option—switching from dairying to grass-fed beef—offers a practical and profitable alternative. Former dairy farmers are well positioned for this transition for several reasons.

- They already have substantial acreage that they own or have under agreement.
- Dairy farmers already know how to make hay and to operate and maintain tractors.
- The biggest advantage for former dairy farmers is that they know cattle. They are accustomed to handling them daily and are tuned in to their condition, behavior, and needs.

An additional benefit for a dairy farmer is that good genetics (semen or breeding stock) for a pasture-only diet are more available in the beef breeds than in the dairy breeds.

A caveat is that a dairy farmer wanting to convert from dairying to finishing grass-fed beef cattle would most likely need to make (or attract) an investment for infrastructure such as a water system, a handling system, and fencing.

The naysayers ignore the rewards of separating cow-calf production and finishing into two operations.

As we have explained, cows need only enough energy for maintenance and nursing a calf, whereas finishing cattle, or *feeders*, need a lot more. So, if the herd is grazed as one unit, the feeders are held back by the cow-calf pairs and don't get the nutrition they need to fatten. These feeders will finish eventually but it will take a lot longer. But on the finishing farm with multiple moves per day, the rate of gain increases dramatically, leading to a shorter and more profitable finishing period.

Look at the numbers:

A 750-pound yearling (12 months) needs another 500 pounds to reach market weight of 1,250 pounds.

He gains 2.5 pounds per day on energy-rich grass, so he'll need an additional 200 days, or 6.5 months, to add those 500 pounds.

Since he was 12 months when he started, the additional 6.5 months will get him finished at 18 to 19 months.

At a slightly slower rate of gain, say 2.25 pounds per day, he would finish in 19 to 20 months.

From Dairy to Beef

Helm and Nancy Nottermann of Snug Valley Farm started milking registered Holsteins in Vermont in the early 1970s, but ten years later, when the price of milk had plummeted, the family realized that to keep their farm operating they had to give up dairy and try something else. At first, they raised and sold registered Holstein replacement heifers to the dairy farms that had not gone out of business as a result of Ronald Reagan's buyout bill in 1985.

Meanwhile, every year the family was also raising five or six grass-fed beef cattle that they sold to family and friends. Eventually, as they gained knowledge and experience with sustainable methods, they expanded that enterprise. Today they raise about 150 head of 100% grass-fed beef cattle. The decision to use regenerative grazing methods was based mainly on practical and financial considerations. In a phone interview, Helm and Nancy's son, Ben, said, "In the beginning the family was unaware of the environmental benefits."

After college, Ben started reading and going to conferences about grass-fed beef production and became actively engaged in the farm in 2007. He is now co-owner and operations manager.

"We always had the rotation of paddocks, but we weren't managing it very intensively. Now I move the cattle two or three times a day," Ben said.

Since we have achieved a three-pound-per-day gain with a cocktail cover crop, the scenario above is certainly achievable.

If calves are born in May, it is quite possible for them to be sent to slaughter before the second winter or the beginning of the nongrowing season, provided they have had a good rate of gain and are sufficiently fat for the marketplace. Harvest at eighteen months would be November. If the calf is to be harvested at twenty months (January) that means the animal needs stored feed at the end of its life, unless there is a cover crop in

A few years ago, the farm switched from Holsteins to cross-bred Angus cattle. "The Holsteins have such a large frame; it took too long to fatten them," Ben explained. The farm buys yearlings from cow-calf producers and raises them for a season and a half before they are ready to market.

Because Snug Valley used to be a dairy farm and there are still dairy farms in the neighborhood, the farm never stopped feeding their cattle dairy-quality hay. "I mow when the dairy farmers mow," Ben said with a laugh. He gets his first cutting in May. Ben's haylage and baleage are good enough to fatten the cattle over the winter; they gain weight at a range of 1.5 to 1.75 pounds a day on the stored feed. Given the high price of grain, Ben is particularly pleased that they are a 100% grass-fed beef operation.

In addition to finishing Snug Valley's cattle, Ben sometimes custom grazes cattle for other farms, and also for Big Picture Beef. Finishing cattle requires time and skill that few people have, so he finds this a good sideline business and may do more custom grazing, especially if he expands the farm's land base. The farm also sells pork and pumpkins. Ben thinks it's important to be diversified and to stay involved with the local community.

the rotation for the late-fall months. (In New England the growing season ends in October.)

Also, positive economics of finishing are partly about the size of the herd being finished. One person can move eight hundred head of cattle to another paddock as easily as that person can move thirty head of cattle. So the more cattle in the finishing herd the lower the cost of labor to move them.

———————

In many businesses, making a profit can mean cutting corners, that is, doing something in a way that saves time or money, though perhaps at the expense of doing it properly. But in the business of raising grass-fed beef cattle, doing things properly—in the interest of the animals' well-being or the soil's health and fertility—usually ends up benefiting the bottom line. In this chapter we have offered specific strategies and practices that grass-fed beef producers can apply either to save money or to make money in a way that fits with your goals for the health of your land and your animals.

If you have good cattle for grass-fed beef and are grazing regeneratively and following a protocol for the care of your animals and your soil, you can build on that strong foundation by experimenting with some of these management ideas. Each year, document what you've done differently and what difference that made. It won't take long for the financial future of your farm or ranch to look brighter.

PART 3

Remaining
Challenges

CHAPTER 8

Cattle into Beef

*Modern industry has de-skilled people and produced what
is probably the most helpless society on the face of the earth.*

—GEORGE McROBIE[1]

O ne thing that both conventional and grass-fed beef producers agree on wholeheartedly is that four giant beef processing companies in the United States are making far too much money at the expense of farmers and ranchers. And the Biden administration agrees.

Beef processors are called *packers*. The business model of these giant packers known as "the Big Four" allows them to control both the price of cattle and the supply of meat to consumers. Frustration with their control of the industry has been brewing among beef cattle producers for years, and now it is boiling over. All of these industry giants are processing feedlot beef, not grass-fed, but the salient issue that could rock the industry is not about the cattle or the meat, but alleged antitrust violations. The chair of the House Agriculture Committee, David Scott, opened an April 2022 hearing on the packers' practices with these remarks:

When a small number of companies control an entire link in the supply chain it makes us more susceptible to shocks and less resilient when black swan events occur. In that vein, consolidation doesn't just hurt ranchers, it also hurts consumers, who face supply bottlenecks, higher prices, and limited choices.[2]

Processing encompasses a number of activities: handling the animal, killing the animal, dismembering the carcass, butchering, and any additional steps prior to and including packaging the beef. But depending on the size of the operation, these tasks can range from artisanal methods implemented on the farm where the animal was raised (with a mobile slaughter unit), to assembly-line work in an industrial plant that slaughters five thousand animals per day.

In this country, there are small, medium, and large packers, plus the four giant corporations that follow.

1. Tyson Foods, out of Springdale, Arkansas, has 25 percent of the market—a slightly larger share than the other three. Tyson's daily slaughter capacity (total of all their US plants) is 28,700 animals and they sell meat products in ninety countries.
2. Cargill Meat Solutions, out of Wichita, Kansas, is one of seventy-five businesses under Cargill Inc., the largest privately held corporation in the United States; it employs 131,000 people in sixty-six countries.
3. JBS Foods USA, based in Greeley, Colorado, bought Swift & Company in 2007, Smithfield Foods in 2008, and Pilgrim's Pride in 2009. In 2008, the US Department of Justice opposed their acquisition of National Beef Packing Company. The parent company, Brazil-based JBS SA, is the largest beef packer in the world, with fifty-four plants on four continents.
4. National Beef Packing Company is the smallest of the four, with just over 10 percent of the US market share and beef sales of $8.5 billion. Their other main product is leather. The parent company is the Brazilian beef producer, Marfrig Global Foods, SA.

Each have billions of dollars in annual sales.[3] Among them, they own twenty-four huge packing facilities around the country and control 85 percent of the finished beef market.

How Processing Has Changed

Of course, butchering wasn't always an assembly-line business dominated by a handful of multinational corporations. To understand how beef producers have found themselves in the financially challenging positions they are in today in relation to the packers, let's take a quick look at significant developments in beef processing from the late nineteenth century to the present:

From the 1860s, with the coming of railroads to Chicago, the Midwest came to be the core of the beef industry, with the Chicago stockyards being the nation's biggest livestock market, and Cincinnati, Omaha, and Kansas City being major meatpacking hubs.[4]

In the 1870s, the use of assembly lines to process beef carcasses became widespread and is said to have inspired Henry Ford to use them for building his automobiles.

By 1900, meatpacking was America's biggest industry, though far less automated and less sanitary than today, as famously described in 1906 by Upton Sinclair in *The Jungle*, a book that drove some regulatory reforms.

After World War II, a series of factors that we outlined in chapter 1—especially grain subsidies—led to bumper crops of corn produced in the Midwest and Great Plains. The problem of what to do with it all led to the proliferation of feedlots that fatten cattle on corn and other grains as grain became—and still is—relatively inexpensive. Cattle from many farms were aggregated in feedlots to spend the final months of their lives eating corn-based rations instead of grass.

In the 1960s, with the shift from railroad transport of meat to refrigerated truck transport, slaughter plants and processing plants were built near the feedlots. Then slaughter and processing began to take place under one roof, and over the years those roofs became larger.

As slaughter and processing became decentralized, business at Chicago's Union Stock Yards declined and the facility closed in 1971.

In the late 1960s, instead of sending sides of beef to stores, the big plants began sending out *boxed beef*, that is, boxes with meat to be butchered further into various cuts: steaks, roasts, and so on. By the 1980s all the supermarkets were receiving their beef this way. Boxed beef also goes to restaurants, institutions, and to intermediate facilities that do custom butchering and packaging.

In the late 1970s, buyouts and mergers ramped up consolidation of meatpacking dramatically.

Fifty years ago, ranchers got over 60 cents of every dollar a consumer spent on beef, compared to about 39 cents today.[5]

According to the USDA, between 1977 and 1997 the percentage of cattle slaughtered in small packing plants (defined as killing fewer than half a million cattle a year) dropped from 84 percent to 20 percent.[6]

Between 1998 and 2018, the number of US cattle producers dropped by 175,000, from 875,000 to 700,000.[7]

Control of the Supply

Size is not the only thing that distinguishes the Big Four from other processors. The giant packers also have a completely different business model. Whereas the smaller packing plants slaughter and process cattle that belong to farmers and ranchers—their whole business is custom processing—the giant packers own all the cattle that they slaughter; they have already purchased every animal that walks into their facilities. Their business model is *vertical integration*, whereby a company gains control of an entire supply chain, from the raw materials to the packaged product. In the United States the Big Four control most of the beef business: from determining the value of the cattle on many farms and ranches, right through slaughter and processing, and to the distribution of packaged meat to stores, restaurants, and institutions.

Often processing companies are invested in feedlots. By the 1990s, cattle breeding was planned with feedlot profits in mind. The industry wanted large-framed animals that could turn concentrated, corn-based rations into muscle meat, and plenty of it. Now the feedlots are filled with these behemoths. The corn-based feedlot system continues to dominate the industry.

Once the live animals have become beef in the processing facility, the farmers, ranchers, or aggregators who sold the animals to the packer generally become customers of the packer; they can buy back the parts of their previously owned cattle—steaks, roasts, livers—whatever parts they can use or sell to their own customers. They get a break on the price of whatever they buy back that is based on the number of head they sold to the plant—that's where they achieve economy of scale. Also, they don't have to pay for any additional costs such as an *E. coli* test, which otherwise costs $75 per head. (The giant packers process multiple tests at lightning speed.) A big appeal for the farmers and ranchers is that they can leave the parts that they can't sell for the packer to market; they buy back only what they can sell to their own customers.

This brings us to an important concept in the beef industry known as *balancing the carcass*. The goal of a meat businesses is to sell every part of an animal, ideally at a profit. But with beef there is much more demand for the steaks (strips, ribs, and tenderloins) than for ground beef. This is problematic because a large proportion of a carcass gets ground. So even though the steaks sell for more per pound, a company needs to sell the ground meat as well as the steaks in order to be profitable—and it can be difficult to sell all that ground meat. Many conventional producers have been willing to sell their cattle to one of the Big Four even if they are not happy with the price they get, because they can leave most of the ground meat, and anything else they don't want, with the packer. This type of arrangement might appeal to grass-fed beef producers as well—if their businesses were large enough to participate.

Vanishing Skills

While the big processors were busy consolidating in the 1970s, they were also perfecting their processing; many technological advances were made in that decade. After the animal is slaughtered, several processes must take place before the meat is delivered to the grocery store, restaurant, or institution. To understand what consolidation has meant to people who work in the processing sector of the meat industry, it's helpful to know what happens in packing plants and what happens elsewhere.

Immediately after slaughter, the hide, head, legs, and organs are removed. The carcass then becomes two "sides" of beef, which need to be cooled. In Big Four plants, the sides might be in the cooler for thirty-six hours.[8]

Then the meat is cut off the sides into the big chunks of meat called *primals* or *subprimals*. (The industry term for this process is *breaking the carcass* —although the carcass is already split into the two sides and the meat is cut, not broken off.) As we've noted, in large plants the disassembly of the carcass is done by people along an assembly line. Every person on the line has a station, and repeats the same cut all day, as chunks of meat, hung from a moving chain, pass their station. (The chain moves quickly and the pieces of meat arrive constantly.) Nobody on the line learns how to do the whole job of transforming a side of beef into primals and subprimals.

In smaller plants there are no assembly lines; the job of cutting the meat into primals and subprimals is done by a team, and it is much more of a craft process that requires a higher level of skill. But the smaller plants don't always have enough skilled workers to break carcasses efficiently and at a

Aging Beef

For beef coming out of the Big Four plants, *aging* happens to the meat after it is packaged in Cryovac plastic and boxed; this is called *wet aging*. Boxed, wet-aged beef can be safely transported or stored for weeks with no deterioration. In contrast, smaller plants typically hang carcasses for three to fourteen days or more. This process is called *dry aging* and is costly because it requires refrigerated storage. The desiccation (thorough drying) of the meat means that more must be trimmed off and thus wasted after the aged carcass is *broken* into the two sides. But many grass-finished beef producers require dry aging of sides because it mitigates tenderness issues. We have noted that tenderness is largely a function of the choice of breed and grazing management.

Both wet aging and dry aging allow the meat to become more tender. Some vendors further dry-age certain high-value primal cuts like the rib and strip for as long as thirty to forty-five days in a climate-controlled environment.

reasonable cost. In that case, the process may depend on only one or two people with adequate hand skills, and therefore becomes a bottleneck for additional throughput at the plant.

How skilled does one have to be to cut a side of beef into primals efficiently? Highly skilled. Ridge witnessed a supermarket butcher—whose skills were cutting primals into steaks and roasts—attempt to cut up a side into primals; it took him four and a half hours. At the other end of the skills continuum, Ridge saw an older man at a plant in Maine do the same job in thirty minutes.

Most beef processing in the United States takes place at one of the Big Four's multiple packing houses; all meat leaves these plants in a box that contains the primals and subprimals. From the packing house, the boxed beef goes to an intermediate facility that performs a variety of butchering and packaging tasks. This facility cuts the primals into specific cuts. The industry term for this process is *portion control*, which means cutting the meat to exact dimensions.

Among these intermediate processors—not to be confused with packing houses that slaughter as well as process—there are some high-end purveyors that supply white tablecloth restaurants with exactly what they need: for example, a whole case of eight-ounce tenderloins or twelve-ounce rib-eye steaks. Most of these purveyors receive their meat as boxed beef and have a staff of skilled butchers who can use a knife to cut exactly the right weight steak, which by law must be within one ounce of accuracy. But all over the meat industry, this is an increasingly vanishing skill, and when the cutting is done by an inadequately trained person, the result is poor yields of saleable meat from the most highly valued parts of the carcass.

Because skilled butchers have become a rarity, many purveyors have turned to very expensive, fully automated machines that measure the primal with a laser and then cut it accurately into steaks in the blink of an eye. These machines are advertised as capable of making a thousand fixed-weight cuts per minute.

Once the meat has been cut to specification it is sent to the wholesale customer (for example, a grocery store, a deli, or an expensive restaurant), either packaged individually or in cases containing multiple pieces of meat, depending on the customer's needs.

In contrast to the giant plants, smaller slaughtering facilities do all the processing at the same location where the animal is slaughtered. After the primals and subprimals are cut from the sides, they are further cut into steaks, roasts, and so on and then packaged and labeled according to the customer's specifications; small processors make most of their profit from this final custom processing. But the portion-control skills are not always adequate. If the butcher repeatedly cuts slightly over the specified weight—of a steak, for example—so that a small piece needs to be cut off and put aside, the result may be a low yield of steaks from that primal. At small plants we have experienced yields as low as 50 percent of the primal, whereas yields by skilled butchers are 80 to 90 percent of the primal. All of the high-value meat that does not become steaks is referred to as *block trim*. Because it was handled by the butcher, block trim has a higher possibility of having a contaminant such as *E. coli* and therefore has greatly reduced value.

When US Cities Had Meat Districts

Coexisting with the meatpacking industry that arose in the late 1880s was another model of getting meat from farm to plate that flourished until the 1970s, when the beef industry consolidated. In this parallel model, cattle were killed in rural areas and the carcasses cut in half and transported to butcher shops in the cities. Beginning in the mid-nineteenth century, every city in the United States had a bustling meat district consisting of dozens of small butcher shops. In New York City's meat district, there were hundreds of meat businesses.

By the mid-twentieth century, the carcasses came into the meat districts in "rail trucks"—not railroad cars, but refrigerated trucks with rails mounted to the ceiling. (Before refrigeration, blocks of ice were used to cool and maintain the meat in these trucks.) The half-carcasses that hung from the rails were referred to as "swinging sides." The city butcher shops all had porches, which were likewise equipped with a rail system to accept delivery of the sides. Jeremiah Moss, author of *Vanishing New York*, writes, "Under metal awnings, sides of beef hung on hooks, dripping blood and fat onto the sidewalks, where men in red-smeared white smocks toiled in the pre-dawn dark."[9] The sides were rolled into the shops on the rail, where butchers would break them into primals and subprimals, and eventually cut them into steaks, roasts, ground—whatever was needed by their store and restaurant customers throughout the city.

But when processing plants began to send boxed beef directly to stores and restaurants—businesses that used to patronize a local butcher or employ local butchers—urban meat districts all over the nation declined.

According to Jeremiah Moss, in 1974 New York's meat district still had 160 meat businesses, down from perhaps 250 in its heyday. Now only a handful exist. The majority of the historic buildings in New York's meat district have been turned into high-end houses, boutiques, and restaurants.

Historically, butchers in grocery stores or stand-alone butcher shops did the further processing of primals. In some grocery stores, butchers still perform these tasks. But the vast majority of stores receive their meat pre-packaged, and the meat department personnel in these stores mainly keep the meat cases supplied and assist customers. Often, they grind fresh meat and make hamburger patties, too.

In March 2019, Tufts University published a study of the processing industry in New England and New York. Eighty-two percent of the sixty-two smaller USDA red meat plants in New England and New York participated in the survey (with an average of eleven employees per plant). Three-quarters of these (74 percent) identified "access to qualified labor" as their biggest industry constraint. In addition to lack of skills, one of the perceived deficits of the workforce was "lack of understanding regarding the industry's role—and hence the worker's role—in sustaining the local and national food system.[10]

Small Is . . . Challenging

Ever since the 1973 publication of *Small Is Beautiful*, E. F. Schumacher's language and insights have sparked dialogues about *human scale* and *appropriate technology*. So how does the size of the processing company play out for farmers, ranchers, or aggregators?

For example, how big does a beef producer have to be to realize an economy of scale for processing? Most of the plants in the United States are not large enough to offer a discount for volume, and the four giant processors are only interested in working with a customer that brings them at least a truckload of cattle every week (thirty-eight head)—or even five truck loads. So, small-to-medium-sized grass-fed beef producers or aggregators take their cattle to small-to medium-sized packers and therefore do not get a break for volume. The high cost of processing per animal at those plants may mean that producers must sell their beef at a price that is too high for the customers they would like to reach. For aggregators, this might mean that they can't offer farmers/ranchers a fair return for the grass-fed cattle they produce.

Another challenge for small-to-medium-sized grass-fed beef businesses may be finding a packing plant that can provide a reliably high standard of service. If the packer makes an error—such as mislabeling meat packages or sending the meat out in a refrigerated truck at the wrong temperature— that one mistake could cost more than a small business can afford to lose. The giant packing plants, on the other hand, seem to run like well-oiled machines. Inefficient systems, equipment, or personnel are promptly jettisoned—though that might be changing as packing-plant labor is getting harder to find.

Currently, packers are coping with a shortage of labor for several reasons: absences due to the COVID-19 pandemic, immigration policy changes made under the Trump administration, and competition from warehousing jobs, where workers do not have to cope with a cold workplace. A recent article in *Meatingplace* magazine about packing plant working conditions noted, "Americans don't want those jobs."[11]

So that's another consideration. Should people spend their days working on a processing assembly line as though they are a replaceable part of the machinery? It may be an efficient way to process meat, but the repetitive nature of the work means that some of those people may be developing carpal tunnel syndrome, for example. Yes, the giant facility will have someone on staff whose whole job is to minimize and mitigate ergonomic problems. And no, such expertise would not be available for workers in a small plant. But would a plant *need* such expertise if the work were planned so that no one person was making repetitive motions all day? We have noted that in a smaller plant the carcass is broken down by a team, and is more of a craft process.

What about economic justice? In today's beef industry, the giant packing companies are the profit makers. When sales are high, profits do not go back to workers on the assembly line or to farmers and ranchers; processing becomes even more profitable for the processing companies. With the onset of the COVID-19 pandemic, the giant packing companies increased their profits by 300 percent, according to a 2022 release by the White House.[12]

In short, grass-fed cattle producers and aggregators in some regions (for example, the Northeast) are challenged to find affordable slaughter and processing services that fit well with everything they would like their business to do and to be. This problem must be addressed because adequate slaughter and processing capacity in every region is important for food security and resilience.

Another topic relating to the scale of the processing enterprise is the handling or disposition of the *offal* (sometimes called the *drop*). These are the parts of the slaughtered animal beyond the muscle meats, including the hide, liver, heart, tongue, glands, bones, and so forth. Some of these items are removed right after slaughter (that is, taken out of the animal when it is cut in half) and other items, such as the bones, are removed later.

The offal is a profit center for big packers because they are able to sell every item: the edible, the inedible, and the parts that are sometimes eaten, perhaps as a delicacy in certain cultures. The hide becomes leather, the bones become broth, the intestines become sausage casings, and so on. The Big Four can succeed in marketing the offal because they kill enough cattle to assemble huge quantities of diverse saleable parts. The offal is profitable enough that the Big Four typically pass on a *drop credit* to the producers/ customers who have sold them the animals and are buying back parts.

But smaller plants don't have enough volume of these items to market them; buyers of these parts are not interested in handling small quantities. Therefore, the offal is a liability to smaller processing facilities, and many of them charge farmers and ranchers a fee to dispose of it. Sometimes even the hides are trashed or composted rather than sold for leather because leather companies prefer to buy all their hides from a small number of large companies rather than from multiple sources.

Of course, anyone interested in sustainable production and waste reduction wants the entire animal used. And the disposal of the offal at smaller facilities is concerning beyond economic considerations for people who feel that the death of the animal should result in the greatest possible good. While the term "highest and best use" has developed in the context of real estate, it is also the concept behind the hierarchy of *reduce, reuse, recycle, or compost*. Clearly landfilling, incinerating, or even composting animal parts is a waste that should be reduced by making full use of offal.

But as we've indicated, whether or not the whole animal is used depends on the size of the processing company. Currently only the giant packers are able to market all the parts of the slaughtered animal, thereby passing them on to a wide variety of uses. Could the components of the offal from several smaller facilities be aggregated for marketing? This would depend in part on the perishability of each component, but might be possible within a region.

THE LEARNING CURVE

Using the Whole Animal in France

In 2002, I organized a group of Northeast agricultural profession-als to travel to Montmorillon, France, to tour a facility run by the Alliance Pastorale, an organization of farmers with a form and function unlike any in the United States.

The alliance is a huge collaborative of French sheep farmers, representing twenty-two thousand farms, that is funded by the sale of farming supplies. The alliance also provides veterinary support, marketing information, production advice, a print shop, an insurance company, and other farmer services. Another closely affiliated group conducts artificial insemination, and yet another handles slaughter and processing at the Sodem, a slaughterhouse. The sheep are kept segregated by farm until the animals have been slaughtered and the carcasses are on the hook; that's when a sale takes place, with the value of each farm's animals determined by the quality of its carcasses.

Every year the Sodem slaughters approximately five hundred thousand sheep. The hides are pulled for processing and every part of the animal is salvaged for use. They are able to market even the smallest gland—to pharmaceutical companies—because of the tre-mendous scale of the operation and consequent volume of each part that can be gleaned.

At the Sodem, I saw various rooms set up to process these parts. In one room people harvested brains, which they packed in six-pack containers for human consumption; in another room, intestines des-tined for sausage casings were washed; in another, feet were processed into gelatin products; hides were salted in a large cement mixer–type device and then packaged on pallets for export. And so on.

In short, all the parts, as well as the meat, had an economic destination. The scale of the operation made this complete utili-zation economically feasible.

—*Ridge*

Shake-up?

In the last few years, several events have brought public attention to the meat industry and have raised alarm about related issues: the nation's dependence on a small number of huge companies, workers' health and safety, and the unfair allocation of profits.

In 2019, a large Tyson Foods plant in Kansas closed for four months following a fire.

In 2020, COVID-19 caused sickness and death in meatpacking plants, which resulted in temporary closures and meat shortages.

In 2021, JBS Foods detected a ransomware attack on its systems that temporarily closed its US beef plants.[13]

All of these occurrences caused temporary shortages, which brought meat prices up. This was a boon to the processers, who control the system, but not to the farmers and ranchers who raise the cattle. News of packers' sky-high profits infuriated many beef producers, and their outrage received media attention.

Damon Watson, an Oklahoma rancher, said in an online interview, "Most people have options if you're selling something. For farmers and ranchers, you get told by the packers what you're going to get for it and you better hope you're happy with it."[14]

As a result of pressure from many quarters and promises made in the last presidential campaign, in January 2022, the Biden administration issued a release that included these announcements: (1) There will be strong enforcement of an existing federal law intended to protect farmers and ranchers from unfair trade practices, and new, stronger rules under the Packers and Stockyards Act, an existing law designed to combat abuses by the meatpackers and processors. (2) Additionally, federal money has been designated for proposals to build small processing facilities, with the first phase to fund fifteen of them.[15]

This type of relief effort has been tried in the past with little success, and it remains to be seen if there is sufficient commitment to change the entrenched system. And while there is considerable support for breaking up the consolidation of the meat industry and increasing the number of smaller processing facilities, these changes in themselves would not necessarily bring about an increase in regenerative grazing and grass-fed beef.

But one current circumstance that might call attention to the advantages of regenerative grazing over corn-based feedlots is the effort on the part of Western countries to end reliance on Russian oil and gas since the Russian invasion of Ukraine. This conflict has brought about a new urgency to decrease use of fossil fuel. Whereas growing corn to feed beef cattle entails heavy use of fossil fuel in both the manufacture of fertilizer and in the cultivation of cornfields, regenerative grazing is a low-energy methodology.

Perhaps a perfect storm of current events, political will, and public awareness will bring the profound changes we need in beef production as well as processing in order to revive our farmlands and rural economies.

CHAPTER 9

Public Awareness, Public Policies

Eating is an agricultural act.
—WENDELL BERRY[1]

I n this book we have extolled the many benefits of grass-fed beef production such as soil health and fertility, carbon sequestration, biodiversity, and a healthy and delicious protein food, to name a few. And we've explained how these good things come about. How regenerative grazing fosters plant growth, microbial activity, and other natural systems. How care of the soil contributes to the health of the cattle as well as the beef producer's bottom line.

So, what's the downside? Why hasn't regenerative grazing already transformed the landscape? Why is feedlot beef still on the menu?

For years, people have heard that conventional cattle production is bad for the environment; most people do not know that a different way to produce beef has been developed. They cannot imagine how grazing cattle can combat climate change. They have not heard that grass-fed beef is a health food—just the opposite.

But getting updated information to the public is only part of the challenge. In this final chapter we comment on some policies, proposals, and research and marketing trends that are holding back the widespread adoption of regenerative grazing for 100% grass-fed beef and all the environmental and health benefits it represents. We also suggest things all of us can do to bring about the changes we need in agriculture, not only to produce healthy food, but also to stabilize the climate.

Ending Agricultural Subsidies for Corn

We've already delineated the harms to the environment caused by vast monocultures of corn grown in the United States to feed cattle. We've also touched on the harm this feed does to the animals themselves, whose bodies are not made for digesting grains. But perhaps the most important thing to know about this crop is that farmers in the United States would not be growing corn for cattle if our tax dollars did not subsidize it; without the subsidies, cattle would be fattened on grass, because that would be the low-cost production mode.

The term *agricultural subsidies* refers to various financial benefits that the federal government offers farmers who produce commodity crops such as corn, soy, wheat, and cotton. The benefits range from disaster payments to crop insurance to loans. In theory, subsidies help reduce risks that farmers face from weather, commodities markets, and disruptions in demand. Subsidies are a central component of the Farm Bill produced by Congress roughly every five years. While there are diverse opinions about the best way to shield our food producers from catastrophic losses, there is general agreement that the biggest beneficiaries of the current subsidy system are giant agribusinesses.

According to the Environmental Working Group (EWG), most of the financial support goes to the largest and most financially secure growers of commodity crops. Producers of meat, fruits, and vegetables are almost completely left out, though they can sign up for subsidized crop insurance and often do receive federal disaster payments.[2] EWG claims that during the Trump administration, between 2017 and 2020, not only did the total dollar amount of subsidies skyrocket from just over $4 billion to more than $20 billion, the subsidy programs also became more dramatically skewed toward the wealthiest farmers and landowners, with the richest 1 percent of farms receiving almost one-fourth of the funds in 2019, and the richest 10 percent receiving two-thirds of the funds.[3] According to *Forbes*, "Between years 2015 and 2017, more than $626 million flowed to recipients in America's urban areas—cities with over a quarter million residents and no farms."[4]

The National Family Farm Coalition feels that subsidies are helpful to their members, but are inadequate. According to their website, "If a floor price is like a minimum wage, subsidies and insurance are more like food stamps: a critical lifeline, but not enough to replace basic economic justice."[5]

Subsidies for corn—the most abundant crop in the United States—have far surpassed those for any other crop.[6] A decade ago, political historian Heather Cox Richardson listed several impacts of cheap corn that are even more of a problem today.[7] She noted that government subsidies for corn have cemented the power of agribusiness; most subsidy money goes to the big growers that can exercise huge economies of scale. Cheap corn has also changed what we eat, and indeed the ubiquity of corn may be an important factor in our nation's epidemic of obesity. Corn is everywhere in American supermarkets as high-fructose corn syrup in soda and processed foods. It is also in less obvious places, namely in the feedlot beef in the meat case. Richardson speculated that corn may also have changed demographics in Mexico, because heavily subsidized US corn has combined with NAFTA (the trade pact that governs commerce among the US, Mexico, and Canada) to force Mexican farmers to move from rural areas into cites and perhaps to the US to find work on farms here.

Corn and much of our food is inexpensive not only because of subsidies but also because of farming practices that offer high-volume production in the short term, but have damaging long-term consequences: monoculture plantings; the applications of herbicides, pesticides, and chemical fertilizers; and plowing, which causes oxidation of carbon, thus worsening climate change. The folly of using our tax dollars to support corn production couldn't be clearer.

Is there a way to turn this situation around? The Union of Concerned Scientists, in a 2017 report titled "Turning Soils into Sponges," suggested modifications to the federal crop insurance program that would protect soil against floods and droughts.[8] These policy changes would address some of the egregious problems caused by the industrial production of corn and soy and other commodity crops:

- incorporating soil quality and management metrics into insurance premium calculations
- improving enforcement of minimum conservation requirements for policyholders
- incorporating ecological principles and encouraging diversified farms by promoting whole farm insurance policies rather than insuring one crop at a time

As a corrective to the harm caused by large-scale corn production, Ronnie Cummins, the author of *Grassroots Rising: A Call to Action on Farming, Climate, and the Green New Deal*, weighs in on the role of government in agriculture with policy suggestions that are similar to those that Franklin Roosevelt successfully implemented in the 1930s—only these are measures to scale up regenerative food and farming production, including grass-fed beef. They include subsidies, tax incentives, minimum crop price guarantees (parity pricing), and supply management to keep supply balanced with need so that prices are stable.[9]

Restoring a Country-of-Origin Labeling Requirement

The Biden administration has the opportunity to boost rural economies all over the nation simply by requiring that meat labels in the grocery stores tell it like it is.

Currently, the labels on packages of grass-fed beef that say "Product of the USA" do *not* mean that the animal was born, raised, or slaughtered in the United States. This labeling only means that the beef was handled here, that it was cut into smaller pieces or ground into hamburger or otherwise minimally processed and packaged here. In fact, "Product of the USA" on a package almost certainly means that the grass-fed beef therein did *not* come from the United States, even if the package is adorned with American flags. Most likely the meat was imported from Australia, Uruguay, or New Zealand.

Because deceptive labeling on packages of imported grass-fed beef leads shoppers to believe that they are buying American beef, it allows companies exporting meat from overseas to cash in on the desires of many Americans to support US companies. Currently this marketing strategy is perfectly legal, but that was not always the case. The short-lived Country of Origin Labeling requirement (COOL), which required retailers to notify customers where beef came from, went into effect in 2009 but was repealed in 2016.[10]

What difference does it make where grass-fed beef cattle were raised? Customers want to know the country of origin for at least two reasons:

- Many want their food dollars to support the grass-fed beef movement in the United States. Ideally, they would like to buy grass-fed beef in the region where they live, so that the money spent will benefit the local or regional rural economy, and support the farmland and farm families close to home.

• Increasingly, people are associating 100% grass-fed beef with methods that are good for the environment and combat climate change by storing carbon in the ground.

Some people might think, "I want grass-fed beef produced with environmentally friendly methods, regardless of where on the planet these practices are happening." Unfortunately, imported grass-fed beef is generally not produced by regenerative grazing; therefore, buying it will not support the environmental benefits that are associated with grass-fed beef. Even if the grass-fed animals overseas are not fed grain, the conventional grazing method used (which we have described as *continuous grazing*) will not build topsoil, store carbon, or protect against droughts and floods.

Meanwhile, it is surprisingly inexpensive to ship beef across oceans, and therefore US grass-fed beef producers are being undercut by foreign competitors. While US grass-fed is only a small part of the grass-fed market, if the Country of Origin Labeling requirement were reenacted and customers knew where their beef was coming from, we could expect a resurgence in the domestic grass-fed beef market.

So, what is the remedy for this situation? Action at the federal level. On January 3, 2022, the Biden administration issued a release pledging to "issue new 'Product of USA' labeling rules so that consumers can better understand where their meat comes from."[11]

We'll see.

Laboratory Beef

The term *lab meat* can refer to two entirely different food products: either (1) the plant-based burger patties that mimic beef, and came into the marketplace in 2018, or (2) a process under development to create muscle meat from stem cells taken from a cow's body, the goal being to produce meat in large quantities. Both of these ultraprocessed food products—one already marketed, and one only conceptualized—are pitched to the public as avoiding the problems associated with feedlot beef. Given that lab meat has attracted massive investment, let's take a closer look at what these products offer.

Fake Meat

Plant-based alternatives to ground beef (often called *fake meat*) are designed to look, taste, and smell like hamburger patties. They are marketed to vegetarians or *flexitarians*, that is, people who want to minimize their meat consumption for whatever reason. Ingredients of plant-based burgers vary, but none of them include whole foods. All the brands are based on chemically extracted plant protein (from soy, peas, potatoes, and more, depending on the brand), fats, flavorings, and a long string of additional ingredients, many of them unfamiliar to us. FoodPrint, a website dedicated to exploring food production practices, notes a problem with determining exactly what is in these products:

> With all ultra-processed ingredients, the chemical processes used to refine, stabilize and package them don't appear on ingredient lists, even when these processes—everything from the hexane used to extract flavors to the BPA used in packaging—can leave behind trace residues.[12]

Regarding general health outcomes of eating plant-based burgers, a 2020 report by physiologist Stephan van Vliet and others finds them problematic as a meat substitute:

> The mimicking of animal foods using isolated plant proteins, fats, vitamins, and minerals likely underestimates the true nutritional complexity of whole foods in their natural state, which contain hundreds to thousands of nutrients that impact human health. Novel plant-based meat alternatives should arguably be treated as meat alternatives in terms of sensory experience, but not as true meat replacements in terms of nutrition.[13]

One big concern with lab burgers is the harm caused by the industrial production of their plant ingredients. As discussed in chapter 2, the industrial mode of crop production entails monoculture plantings, chemical fertilizers and biocides, fossil-fueled equipment, tillage, bare ground that exposes soil carbon to oxidation, and death to wildlife from farm machinery and habitat destruction. In short, industrial vegetable production is a poor substitute for industrial meat production.

Regarding the harm to the climate that the fake meat brands claim to alleviate, life-cycle analyses show that in comparison to 100% grass-fed beef, the plant-based burgers have the higher carbon footprint.[14] And yet, "plant-based" continues to be a rallying cry for these products and many others. Most people don't know that conventional vegetable production contributes to climate change, whereas 100% grass-fed beef cattle produce a net climate benefit because of the carbon sequestered by regenerative grazing.

Who stands to gain the most if alternative meat is ultimately successful? Perhaps surprisingly, a growing number of alternative-meat brands are owned and operated by major meat-packers like Tyson, JBS, and Cargill.[15] Cargill has reportedly projected that plant-based foods might cut into their own beef sales, but the company is expanding its own soybean processing capacity to supply both livestock feed and inputs for meat alternatives; both use the same plant inputs.

Cellular Meat

Meat created from stem cells sounds like science fiction, but in fact teams of scientists are hard at work in laboratories attempting to create muscle and fat tissue outside an animal's body. At the Kaplan Lab at Tufts University, a research team is devoted to isolating stem cells from cows, proliferating those cells, and turning them into muscle fibers.[16] During the first phase, cells simply divide; in the second phase they stop dividing and start forming different cells, leading to function-specific tissues and organs. Then the cells produce the proteins required for a muscle to contract. The goal is to ramp up the procedure to the level of food production. The full-scale vision is called *cellular agriculture*.

Why would anyone want to replace animal meat with lab meat? According to a *Tufts* publication, the motivation is "to reduce our dependence on animal agriculture and the heavy toll of meat production on the environment, animal welfare, and public health."[17]

The same article says, "Proponents of cellular agriculture believe that it could reduce land, water, and chemical inputs and minimize greenhouse gas emissions." A PhD student who works in the Tufts lab points out that with cultured meat, "You don't have the pollutants just running off into the waterways."[18]

But is it true that cultured meat production would avoid negative environmental impact? We don't believe anyone is prepared to answer that question. Our understanding is that energy-intensive bioreactors are needed to produce stem cells, and the cell medium that grows the stem cells is soy, corn, or wheat.[19] We're already familiar with the environmental and economic impacts of commodity crop production in the United States, and we want to turn that situation around. What other material will be in the media? How much waste will be generated? How will it be discarded?

Note that the environmental arguments for cellular agriculture are much the same as our arguments for regenerative grazing. Of course, producing beef regeneratively in pastures on grass is far simpler and less expensive than developing stem cell technology, and in addition, regenerative grazing builds soil health and fertility, protects against droughts and floods, and sequesters carbon. Perhaps lab meat advocates are not fully aware of this grazing methodology and the peer-reviewed documentation of its benefits.

Another key question without an answer at this time: Will this high-tech approach to providing nourishment concentrate food production and distribution in the hands of a few—those with protected intellectual property and patented technology?

Research follows funding, and the movement for scaling up lab meat production is attracting billions of dollars from private investors (including Bill Gates and Richard Branson), agribusinesses, and from the federal government.[20] The Kaplan Lab at Tufts received $10 million in funding from a USDA grant to optimize its cellular agriculture efforts.[21] The grant will allow Tufts to create the National Institute for Cellular Agriculture, the first government-funded protein research center. One of their goals will be preparing a workforce for the new industry.

Regarding potential displacement of the nation's farmers and ranchers by the new industry, Sean Cash, an associate professor at the Tufts' Friedman School of Nutrition Science and Policy, says, "Creative destruction . . . is always a part of new technologies."[22]

Risky Climate Proposals

Continuing with the topic of costly and unproven technologies, some recent proposals for combatting climate change are troubling. For example, the following high-tech ideas for modifying the climate have been circulating recently, and some of them have funding:

- refreezing the North and South Poles by brightening the clouds above them by spraying tiny droplets of salt to assist the clouds in reflecting radiation back into space[23]
- fertilizing the oceans to encourage the growth of plant matter and algae, which could absorb more CO_2[24]
- injecting captured CO_2 from industry into the deepest parts of oceans, where "most of it would remain isolated from the atmosphere for centuries"[25]

Regarding risks from the third idea, *IPCC Special Report on Carbon Dioxide Capture and Storage*—the same publication that presents the proposal and gives it credence—acknowledges that nobody knows what the impacts of releasing CO_2 into the depths of the ocean would be:

Overall, there is limited knowledge of deep-sea population and community structure and of deep-sea ecological interactions. . . . Thus the sensitivities of deep ocean ecosystems to intentional carbon storage and the effects on possibly unidentified goods and services that they may provide remain largely unknown.[26]

We're not opposed to technology, and we do believe that combatting climate change calls for a multipronged effort. But will those we have entrusted with decisions about the future of human life on Earth also guard against irreparable harm? Often harm occurs as a result of *reductionist* thinking: when scientists—or politicians—focus on one part of a problem and lose sight of the interconnectedness within the whole context.

Because this way of thinking can be dangerous, we often look to biology as a reliable guide to judge which technologies are appropriate. Biological systems were operating long before the first humans appeared, but only recently have people developed a rudimentary understanding of how they actually work. To use two examples of recent discoveries that we've referenced in this book: (1) it was only in the 1990s that the organic material glomalin was identified as a critically important structure in healthy soil functioning, and (2) it was only in this millennium that soil scientists and ecologists have verified that by fostering the activities of soil microorganisms, we can dramatically increase both vegetative growth in farm fields and storage of carbon in the soil. Perhaps instead of relying on unproven technologies to address our climate crises, we should build on discoveries of the last thirty years about how biological processes work—and how we can work with them instead of against them.

Let us pause before we proceed to interfere with the most basic interdependent components of our world: earth, air, water, and biological resources. High-tech schemes to control the Earth's climate would come with enormous expense and possibly disastrous environmental outcomes that aren't being discussed and perhaps haven't been imagined. And while we are distracted by proposals with science fiction allure, we are ignoring simpler, safer, and tried-and-proven ways to stabilize the climate that work with existing natural systems. Salient among these solutions is the widespread regenerative grazing of ruminants to foster carbon sequestration, a scenario that reflects an earlier, but not-so-distant era in the natural history of the Earth.

Regenerative grazing of existing grasslands is a feasible strategy that can be implemented quickly, with low risks, and with multiple benefits in addition to stabilizing the climate. For more than twenty-five years a growing number of farmers and ranchers all around North America and in other parts of the world have increased carbon storage *in the ground* largely by (1) forgoing tillage, chemical fertilizers, and biocides, and (2) grazing cattle in a rotation that fosters robust populations of soil microbes, whose underground activities enhance plant growth and store carbon.

Too simple?

Consider what our country could have accomplished in the last twenty-five years if regenerative grazing had been fully supported and implemented as an alternative to the conventional agricultural practices that have failed to safeguard our resources. What progress would we have made if the federal government had pulled back subsidies for corn and supported regenerative grazing instead? What would our domestic production of grass-fed beef be if we had not allowed inexpensive imports to be labeled as Products of the USA?

If the promise of grass-fed beef and regenerative grazing had been understood, supported, and widely adopted twenty-five years ago, there is no doubt that a long list of outcomes would have been realized at least in part by now:

an increase in soil carbon

fewer climate emissions from agriculture to the atmosphere

increased fertility of farmlands

more soil resilience to droughts and floods

cleaner water

greater health and well-being of livestock

an increase in biodiversity

a revival of rural economies

access to nutritious beef in every region of the country

Many of us would like to see these outcomes. Let's put some of the riskier and more expensive climate change projects on the shelf until we have made a modest investment in regenerative agriculture.

In Conclusion

In the spring of 2022, when our society's attention was focused on front page articles about the Russian invasion of Ukraine, the agricultural press was abuzz with news that the cost of fertilizer was up 300 percent since 2020, and that US farmers were coping with a fertilizer shortage.[27] We were fascinated by the connection between the fertilizer crisis and the stories about economic sanctions against Russia. While the United States does produce fertilizer, we also import some. From Russia.

Kathy Mathers, vice president of public affairs at The Fertilizer Institute, expressed a prevailing view in an online publication, saying, "Corn requires fertilizer and as more farmers turn to growing corn to capture the high commodity price, it is driving global demand for more fertilizer."[28]

In a darkly comic way, the fertilizer articles reflected how oblivious most people seem to be about some of our nation's perilous farming policies and practices.

The full folly of the situation is hard to grasp. Just the production of nitrogen fertilizer—even before putting it on the fields—is a significant source of greenhouse gases, especially nitrous oxide, which is 300 percent more heat-trapping than carbon dioxide.[29] Nevertheless, the United States has been importing nitrogen fertilizer from halfway around the world to grow corn, a crop that is subsidized with our tax dollars and cultivated with harmful impacts that have been documented by our government agencies. In large part that corn goes to cattle feed, which causes the animals to have health problems that are addressed with antibiotics that then show up in their beef, leading to a human health crisis from antibiotic-resistant bacteria.[30] And that's only half the folly. For many years, US farmers have been applying fertilizer to cornfields up and down the Mississippi River watershed, where the runoff has polluted streams that carry the nitrogen to the Mississippi, which shuttles it south, where it enlarges the dead zone in the Gulf of Mexico (measured by the EPA in 2021 as 6,334 square miles[31]), which has been created, in large part, by agricultural inputs to US cornfields. So that's the fertilizer saga: from Russia to the Gulf, repeated annually at considerable expense, causing health problems for cattle and people, and incalculable harm to natural resources. It's not a virtuous cycle.

Nevertheless, some people are worried about the fertilizer shortage. Nobody quoted in the spate of articles we saw questioned the view that our farmlands need more nitrogen fertilizer. Well, almost nobody: one farmer interviewed was not concerned about the shortage or the high price because he had switched to regenerative methods and doesn't use much fertilizer anymore. (Clearly, this fellow is onto something.)

If US agriculturalists replaced feedlots with regenerative grazing and used livestock in their cropping systems to build soil fertility, the United States could increase crop yields without causing harm. Our farmland could continuously rejuvenate itself with minimal off-farm inputs, including fertilizer. And, of course, cattle would be healthier without corn, consumers would be healthier without antibiotics in their beef, and both beef and vegetables would be more nutritious from improved soil health.

But here's another thought: if we *did* need to feed corn to cattle and if we *did* need imported fertilizer to grow corn, the fertilizer shortage would be just one of many out-of-our-control situations making us vulnerable in times of crisis.

Our Failed Food System

Which brings us to a theme of this book. In recent years, potential threats to our US food supplies—from pandemics, ransomware attacks, extreme weather events, terrorism, and wars—have become all too real. The COVID-19 pandemic, which closed numerous meatpacking plants and triggered shipping delays of many farm products, raised alarms about our concentrated and corporate-dominated food system here in the United States. Access to beef is of particular concern because it is perishable and because protein is an essential nutrient in human diets.

And the problems we're facing are not just quantity and availability of food; quality is a problem, as well. Our health is threatened by nutrient-deficient food, chemical inputs to farmland, and antibiotics given to livestock that are present in their meat. Another concern is that some communities don't have access to fresh food; the only food available to them is from fast-food restaurants or stores that carry only highly processed, prepackaged items.

Though controversies continue to swirl around what constitutes the best diet, there is no argument about this fact: health in the United States is below that of other high-income countries in a number of measures, including low life expectancy, high infant mortality, and prevalence of obesity, diabetes, and heart disease.[32] With so much inexpensive food available in the US, we spend relatively little at the grocery store, but pay later in ways we might not consider. These externalities are the hidden costs to our health, our environment, and our society. We pay through our taxes, medical bills, insurance, and decreased quality of life.

Our food system is not working for us and will not work for others who share our planet.

As economies around the world began to emerge from the COVID-19 pandemic, the war in Ukraine caused global food prices to surge, making poor countries especially vulnerable to hunger. Adding to the uncertainties about food supplies is the fact many of Earth's inhabitants are on the move; millions of refugees are fleeing violence, and in the near future many more will be fleeing flooding and other effects of climate change.

These global problems have no simple solution, but there are certain things we must do wherever we live, with an eye toward global realities. Here in the United States, we need to shock-proof our food production and delivery systems by shortening our supply chains. We need to develop ecologically focused, regional supplies: food that is produced, processed, distributed and consumed within a given geographic region of the country.

Developing sources of grass-fed beef in every region of the United States is feasible because regenerative grazing is adaptable to climatic and soil variations, with minimal off-farm inputs required. With the aggregation model we outlined in chapter 6, family farms and ranches and related livestock enterprises can flourish, especially if adequate, appropriately sized slaughter and processing facilities are developed.

Food supplies closer to home—to everyone's home—would require resources and opportunities for small- and medium-sized businesses to succeed. It would also mean prioritizing decent livelihoods and working conditions for people employed in local and regional food systems.

The Whole of Society

A recent study out of the University of Colorado reported that 40.9 percent of *regenerative agriculture* practitioner websites stated that "improving community well-being" is one of the goals of the movement.[33]

As we consider how our food is produced, we should also think about who is empowered by the production and marketing of what we're eating. Are essential laborers benefiting from our food system as well as working in it? While this book has focused on health and environmental impacts of production practices, how we produce food also reflects social injustices and racism endemic in our society. Do some groups in the community face unfair obstacles to owning land or access to job training or to positions with higher wages and responsibilities? Answers can be found in statistics as well as in the stories of people living in the community.

In the United States, 96 percent of farm owner-operators are white.[34] Less than 4 percent of owner-operators are people of color (Black, Asian, Indigenous, Pacific Islander, and those reporting more than one race). This is especially egregious considering that white people who invaded and settled this continent seized the land from Indigenous people; later other white people kidnapped Black people from Africa to labor as slaves of white plantation owners, which made the cotton industry possible and profitable. A series of federal Homestead Acts from 1862 and into the twentieth century were technically open to Black and female applicants, but in fact gave mainly white male settlers and corporations heavily subsidized land.[35]

Regarding jobs in our food system, currently field laborers and assembly-line meat-packers do physically difficult work that is low paying. Historically these jobs have been filled by immigrant labor. Over 80 percent of farm laborers are from Central and South America.[36] But only 6 percent of

An Indigenous Rancher's Story

Kelsey Scott is a Native American co-owner of the grass-fed beef operation DX Beef, on the Sioux tribe's Cheyenne River Reservation. Her family history is the history of Indigenous people in the American West. As white people settled on traditionally Indigenous lands, resulting in violent conflicts, Native American territory was repeatedly reduced to smaller or less desirable holdings throughout the Western states. Scott's great-grandfather had his allotment relocated from a fertile valley (now flooded by a dam) to desert highlands.

Scott and her family are ranching on this desert land but it is less arid now because for years they have been implementing traditional Native practices along with modern technology to make the land fertile and healthy. Scott commented in *Deseret News*, "I say I'm the fourth-generation rancher in my family, but I'm of the 125th generation to help steward the Great Plains. Some of my ancestors helped to evolve this landscape to what it is now. We were a part of the ecosystem ourselves."[37]

owner-operators and tenant operators are Latinx, which is well below their 17 percent representation in the US population.

Gladys Godinez was born in Guatemala and now lives in Lexington, Nebraska, home to JBS Foods and Tyson plants where her parents, now retired, were employed as meat-packers. In an online interview, Godinez said, "Twenty years ago, there was a big wave of immigrants that came to Lexington because a meatpacking facility had opened. They recruited us to this rural community. . . ."[38]

The pandemic brought health concerns to employees of the facility. After former President Trump used the Defense Production Act to ensure that meat processing plants remained open in 2020, Nebraska's governor, Pete Ricketts, refused to close the JBS plant despite requests to do so from public health officials. Godinez said, "During the pandemic, I went to various leaders asking to have Spanish language information throughout the community, because a third of our residents speak [primarily] Spanish; they didn't respond." Not speaking English as a first language is an obstacle to people who need information about their new communities.

As regenerative farming and ranching become widespread, we must ensure that people of color have fair access to the information, training, and financing options that make it possible for people to participate in grass-fed beef production as owners, operators, managers, or skilled workers. If the people in desirable positions in a region's food system do not reflect the population of that region in terms of race and ethnicity, community leaders need to identify and remove the obstacles to opportunities and advancement.

At least two characteristics unique to grass-fed beef production can empower more people who want to be involved in agriculture:

- Regenerative grazing does not require landownership or a huge capital investment to get started, which means that more people from diverse backgrounds can participate in grass-fed beef production as business owners and operators.
- The need for skilled finishers of grass-fed cattle creates a new agrarian niche for people who have the training. Experienced finishers can fatten their own cattle or custom graze for others, on either their own land, someone else's land, or conservation land.

Making Change

Because we all eat, we can begin to change the food system by recognizing our power as consumers and spending our food dollars on healthy food. We can ask our local stores to carry 100% grass-fed beef produced in the United States. Instead of looking for the least expensive food to buy, those who have the means can seek out the high-quality food that our society needs to have available in the future. At the same time, we can boycott foods brought to us by multinational corporations that are pillaging our natural resources for their profit. In the long run, the cost of grass-fed beef and other regeneratively produced foods are a bargain for a livable Earth and the health of generations to come.

But we have to do more than shop wisely. Many of the problems with our food system are political and will not be easy to fix, particularly in an era of partisan politics. A battle cry to tear everything down is unlikely to get adequate support. In this situation, one strategy for effective intervention with elected representatives is to insist on their support for policies and programs that have worked well.

Parity pricing is a good example. The concept is based on the idea that the selling price of a product, such as a bottle of milk or a pound of ground beef, should go up or down according to the costs of the inputs in producing the item. This concept requires supply management, as well. The point is to preserve the real value of the product. Parity pricing has been done before; it served our farmers for more than half a century (until its demise under Reaganomics) and needs to be retrieved from recent history to serve us again in the current time of need. As a letter to the editor from a Pennsylvania farmer noted this spring:

> Parity pricing should not be dismissed. It has worked in the past and will work now. The country is on the edge. Parity pricing will not only fix the agricultural problem but the economy as well. We haven't fixed it before because politicians and special interests want power, money and control. Their goal is to change the country for their benefit, not the benefit of the country or the people.[39]

Although our federal government currently subsidizes corn (among other harmful policies), the USDA does have some good programs in place, and these should be identified and prioritized for funding. For example, the Conservation Reserve Program (CRP) pays a yearly rental to farmers to plant species that will establish land cover to improve water quality, prevent erosion, and reduce loss of wildlife habitat; this program could and should be dramatically expanded, especially as it fits with the goals and implementation of regenerative grazing. Another example of an existing, effective government agency—one that actually goes out into the field to help farmers safeguard resources—is the NRCS (Natural Resources Conservation Services), whose knowledgeable staff can and does assist grass-fed beef producers and other regenerative farmers. Similarly, ATTRA—a program of the National Center for Appropriate Technology (NCAT), which gets a tiny amount of federal

Common Ground in the Partisan Divide

Back when our now middle-aged son was an adolescent, we invited a sixteen-year-old, homeschooled, Nebraskan farm boy to spend three weeks with our family. This was thirty years ago, before our society had settled into the polarities that divide us now, so neither set of parents was prepared for the differences in the two families' politics, religious beliefs, and opinions on social issues; but these differences became apparent immediately. What saved the situation from being a troubling and perhaps frightening predicament for the teenager was his comforting discovery on his first evening in our home that Lynne was reading *Little House on the Prairie* to our young daughter; his mom was reading the same book to his little sister.

Finding commonalities is challenging in the current political climate, but it is still possible. While consulting around the United States and Canada, I've seen the topic of grass-fed beef production unite people from different perspectives: right and left, rural and urban. It is easy for me to avoid overtly political discussions because there is typically enthusiastic agreement about caring for the land and water, the challenges of producing healthy food, and

funding—is a great resource for regenerative farmers and ranchers. In contrast to information funded by the USDA that promotes destructive practices, NCAT's publication, "Building Healthy Pasture Soils" (as one example) is first-rate and contains invaluable information for farmers and ranchers.

Of course, the major problem is the disparity in funding between the programs that have real value—but barely survive administrations that are hostile to conservation—and the programs that promote harmful practices the importance of thriving rural economies to sustain families in the future.

The grass-fed beef ranchers I have visited are excellent conservationists. For example, a Montana grass-fed beef producer has enhanced his ranch's stream habitats to bring about the resurgence of a rare species of trout. Most of his land is under conservation easements. Like most grass-fed beef operations, his ranch ensures soil health with regenerative grazing and cover crops. A core value for this rancher is to leave the land's natural resources in pristine condition for future generations.

On the West Coast, another example of an unlikely person-to-person interaction took place regularly when a large beef cooperative required members to do annual in-store visits—wearing their cowboy hats—to give out meat samples and talk to customers. In San Francisco, this cooperative found that the conversations around the sampling table provided an unexpected, "magical" connection between the city residents and the rural ranchers. They were often of different political persuasions, but over a delicious morsel of healthy food they related to one another as people.

—Ridge

but survive because they are propped up by corporate lobbying. Our government must eliminate the programs that are squandering our natural resources along with our tax dollars, and ramp up support for those that are largely overlooked but highly relevant and effective. Getting the attention and support of decision makers is the challenge.

We all need to insist on major changes via the 2023 Farm Bill. Call on your elected officials to reject our failed food system and prioritize regenerative agriculture. Ask friends to join you in this effort; three people can be an effective advocacy group. Or join an established organization's campaign.[40] Eventually many voices can be a crescendo demanding action for farm families, farmland, and our future.

Confront lawmakers with these basic, well-documented facts:

- Industrial agricultural practices for both meat and vegetable production are wasting fossil fuels and other resources, and driving climate change—*and yet they are supported by public policies and subsidized with public funds.*
- Regenerative agriculture can restore our farmlands, produce an abundance of healthy food, and combat climate change, all with minimum inputs—*but regenerative farmers and ranchers are not supported by government-funded programs.*
- Grass-fed beef production can succeed in every region of the country and help move us toward a healthier and more equitable society—*but success depends on ending subsidies for cattle corn, restoring the country-of-origin labeling law, and funding policies that support regenerative practices.*

Regenerative grazing should have a lead role in a new committment to provide healthy food, heal our ecosystems, revive our rural economies, and stabilize our climate. Equipped with the information in this book and other resources we have referenced, every one of us can support and advance this approach to grazing cattle, which offers multiple benefits to our country and our world. The need is urgent and the methodology is at hand.

Acknowledgments

O ver decades of learning about cattle and regenerative grazing, and eventually founding a grass-fed beef business, we were lucky to have had wonderful support from family, friends, mentors, and other knowledgeable and generous people, including the Arnow family—Josh and Elyse, Talia, Chloe, and Eli—Dan Barber, Sue Beal, Tina Bielenberg, Marc and Cheryl Cesario, Abe Collins, David Couch, Annie Farrell, Doug and Sarah Flack, Roger Fortin, Cary Fowler, Gearld Fry, Tom Garnet, Brooke Henley, Libby Henson, Jock Herron, Nancy Kohlberg, Chuck Lacy, Ken McDowell, Kim Miller, Norm Nick, Scott Phillips, Bill Roberts, Jerry Shinn, Allen Williams, and many others whose help we have appreciated.

We also want to acknowledge people who kindly reviewed chapters of this book or provided other advice, information, or skilled assistance with preparing the manuscript, including Katherine Bourbeau, Gabe Brown, Gwen Broz, David Brule, Bruce Griffin, Newell Isbell Shinn, Susan Middleton, Raymond Lanza-Weil, Gary Partenheimer, Edgar Stewart, Richard Teague, Sue Van Hook, Jim Wagener, and Lisa Walker.

Finally, we are grateful to the skillful staff at Chelsea Green Publishing, and especially our editor, Ben Trollinger.

Resources

Educational Programs

ATTRA
https://attra.ncat.org/wp-content/uploads/tutorials/managed-grazing/introduction/
Online tutorial with phone support, specifically for managed grazing; divided
into topics.

FUTURE HARVEST
https://futureharvest.org/resources/resources-for-farmers/
A variety of hands-on farm programs in the mid-Atlantic states designed to
"advance agriculture that sustains farmers, communities, and the environment."

NEW ENGLAND GRAZING NETWORK (NEGN)
https://www.negrazingnetwork.com/negn-partners/
Events and training opportunities online and in person to increase well-managed
grazing in New England.

DAIRY GRAZING PROJECT
https://dairygrazingproject.org/
Training and support for dairy farmers in the Chesapeake Bay watershed, to
improve grazing knowledge and skills.

PASTURE PROJECT AT THE WALLACE CENTER
https://pastureproject.org/
A wide variety of ongoing programs in the Upper Midwest to advance regenerative grazing, value chains, and racial equity through research, education, and
network building.

WENDELL BERRY FARMING PROGRAM
https://www.sterlingcollege.edu/wendellberry/
Undergraduate farming program in Kentucky (tuition-free, focused on ecological
management and based on the lifework of farmer and writer Wendell Berry).

Books

Dirt to Soil by Gabe Brown (Chelsea Green Publishing, 2018)
The Art and Science of Grazing by Sarah Flack (Chelsea Green Publishing, 2016)
Humane Livestock Handling by Temple Grandin (Storey Publishing, 2008)
Dirt: The Erosion of Civilizations by David R. Montgomery (University of California
    Press, 2012)
What Your Food Ate by David R. Montgomery and Anne Biklé (Norton, 2022)
Farming While Black by Leah Penniman (Chelsea Green Publishing, 2018)

Videos

The Soil Story (3.29 minutes) by Kiss the Ground, narrated by Larry Kopald
Carbon Cowboys (3.44 minutes) by Carbon Nation, narrated by Peter Byck

Notes

Introduction

1. Henry A. Wallace, "The Strength and Quietness of Grass," radio speech delivered by Vice President Henry A. Wallace on June 21, 1940, https://torgprom.blogspot.com/2007/06/strength-and-quietness-of-grass.html.

2. United Nations, "The State of Food Security and Nutrition in the World 2021 (SOFI)," https://sdgs.un.org/events/state-food-security-and-nutrition-world-2021-sofi-33052.

3. Mike Dorning, "Ranchers' Ire at 'Red-Line Level' as Packers Pocket Beef Profits," *Bloomberg*, May 20, 2021, https://www.bloomberg.com/news/articles/2021-05-20/it-s-not-just-shoppers-riled-by-pricey-beef-ranchers-seethe-too.

4. Jennifer Clapp, "Spoiled Milk, Rotten Vegetables, and a Very Broken Food System," *New York Times*, May 8, 2020, https://www.nytimes.com/2020/05/08/opinion/coronavirus-global-food-supply.html.

5. Greg Henderson, "JBS Plants Operational after Cyberattack," *Drovers*, June 2, 2021, https://www.drovers.com/news/industry/jbs-plants-operational-after-cyberattack.

6. "Las Damas Ranch Case Study," *Understanding Ag*, https://understandingag.com.

7. Renee Cheung, Paul McMahon, et al., *Back to Grass: The Market Potential for U.S. Grassfed Beef* (Stone Barns Center for Food & Agriculture, Armonia LLC, Bonterra Partners, and SLM Partners, April 2017): 6, https://www.stonebarnscenter.org/wp-content/uploads/2017/10/Grassfed_Full_v2.pdf.

8. J. Howell, "Can We Do It on Grass Alone? Beef Production and the Unrealized Capacity of Grasslands," Quivira Coalition presentation 2012, http://quiviracoalition.org/images/pdfs/1/5241-Howell_Quivira_2012.pdf.

9. W. Richard Teague, "Forages and Pastures Symposium: Cover Crops in Livestock Production: Whole Systems Approach: Managing Grazing to Restore Soil Health and Farm Livelihoods," *Journal of Animal Science* 96, no. 4 (2018): 1519–30, https://doi.org/10.1093/jas/skx060.

10. Emily Payne, "Dr. Richard Teague: Regenerative Organic Practices 'Clean Up the Act of Agriculture,'" AgFunder Network Partners, June 21, 2019, https://agfundernews.com/dr-richard-teague-regenerative-organic-practices-clean-up-the-act-of-agriculture.

11. Ronald S. Oremland and Charles W. Culbertson, "Importance of Methane-Oxidizing Bacteria in the Methane Budget as Revealed by the Use of a Specific Inhibitor," *Nature* 356, no. 6368 (1992): 421–23, https://doi.org/10.1038/356421a0.

12. Commission on Ecosystem Management, "Dryland Ecosystems," https://www.iucn.org/commissions/commission-ecosystem-management/our-work/cems-specialist-groups/dryland-ecosystems.

13. Rebecca Paredes, "Where Have All the Buffalo Roamed?" *Green Future*, December 30, 2016, https://greenfuture.io/sustainable-living/are-buffalo-extinct/.

14. Action Against Hunger, "World Hunger: Key Facts and Statistics 2022," https://www.actionagainsthunger.org/world-hunger-facts-statistics.

15. W. Richard Teague et al., "The Role of Ruminants in Reducing Agriculture's Carbon Footprint in North America," *Journal of Soil and Water Conservation* 71, no. 2 (2016): 156–64, https://doi.org/10.2489/jswc.71.2.156.

Chapter 1. Regional Resilience

1. Wendell Berry, "Renewing Husbandry," *Orion*, https://orionmagazine.org/article/renewing-husbandry.

2. The White House Briefing Room, "Fact Sheet: The Biden-Harris Action Plan for a Fairer, More Competitive, and More Resilient Meat and Poultry Supply Chain," January 3, 2022, https://www.whitehouse.gov/briefing-room/statements-releases/2022/01/03/fact-sheet-the-biden-harris-action-plan-for-a-fairer-more-competitive-and-more-resilient-meat-and-poultry-supply-chain.

3. Charles A. Taylor, Christopher Boulos, and Douglas Almond, "Livestock Plants and COVID-19 Transmission," *PNAS* 117, no. 50 (November 19, 2020): 31715, https://doi.org/10.1073/pnas.2010115117.

4. Kyle Bagenstose and Sky Chadde, "Trump Executive Order Didn't Stop Meat Plant Closures. Seven More Shut in the Past Week," *USA Today*, May 6, 2020, https://www.usatoday.com/story/news/investigations/2020/05/05/coronavirus-closes-meatpacking-plants-despite-trump-executive-order/5172526002.

5. Gilda V. Bryant, "Grass-Fed Beef Sales Jump During Pandemic," *Progressive Cattle*, July 24, 2020, https://www.progressivecattle.com/topics/management/grass-fed-beef-sales-jump-during-pandemic.

6. Sam Gugino, "Steak Out," *Wine Spectator*, November 2002.

7. Rex Dalton, "First Mad Cow Found in America," *Nature*, December 30, 2003, https://www.nature.com/articles/news031229-3.

8. Renee Cheung, Paul McMahon, et al., *Back to Grass: The Market Potential for U.S. Grassfed Beef* (Stone Barns Center for Food & Agriculture, Armonia LLC, Bonterra Partners, and SLM Partners, April 2017), 16–19, https://www.stonebarnscenter.org/wp-content/uploads/2017/10/Grassfed_Full_v2.pdf.

9. NielsenIQ, "Answers on Demand; Protein Sales, January 2019–November 2021, Processed 11/22/2021."

10. Nielsen, Retail Sales Data. Available at https://www.nielsen.com/us/en/ (accessed July 13, 2020), cited by Stephan van Vliet, Frederick D. Provenza, and Scott L. Kronberg, "Health-Promoting Phytonutrients Are Higher in Grass-Fed Meat and Milk," *Frontiers of Sustainable Food Systems* 4 (2020): 555426, https://doi.org/10.3389/fsufs.2020.555426.

11. Deena Shanker, "Most Grass-Fed Beef Labeled 'Product of USA' Is Imported," *Bloomberg*, May 23, 2019, https://www.bloomberg.com/news/articles/2019 -05-23/most-grass-fed-beef-labeled-product-of-u-s-a-is-imported.

12. Bob Benenson, "Grazing in the Grass Is Growing Fast," *New Hope Network*, April 3, 2019, https://www.newhope.com/market-data-and-analysis/grazing -grass-growing-fast.

13. "Grass-Fed Beef Market Report 2021: Industry Analysis (2017–2020) & Growth Trends and Market Forecasts (2021–2025)—ResearchAndMarkets .com," *Business Wire*, March 25, 2021, https://www.businesswire.com/news /home/20210325005526/en/Global-Grass-Fed-Beef-Market-Report-2021 -Industry-Analysis-2017---2020-Growth-Trends-and-Market-Forecasts-2021 ---2025---ResearchAndMarkets.com.

14. Most of the names of businesses carrying grass-fed beef came from grass-fed beef consultant Allen Williams, in email correspondence with the authors, May 3, 2021.

15. Jennifer Strailey, "Grass-Fed, Wild-Caught Is the Meat of This Story," *Winsight Grocery Business*, July 11, 2019, https://www.winsightgrocerybusiness. com/fresh-food/grass-fed-wild-caught-meat-story (accessed April 28, 2022).

16. R. L. (Bob) Nielsen, "Historical Corn Grain Yields in the U.S.," Corny News Network, August 2021, https://www.agry.purdue.edu/ext/corn/news/timeless/ YieldTrends.html.

17. Economic Research Service, "Cattle & Beef: Sector at a Glance," U.S. Department of Agriculture, November 29, 2021, https://www.ers.usda.gov/topics /animal-products/cattle-beef/sector-at-a-glance.

18. Mike Baker in phone conversation with Ridge Shinn, January 2021.

19. Shanker, "Most Grass-Fed Beef Labeled."

20. Strailey, "Grass-Fed, Wild-Caught."

21. Emily Payne, "Dr. Richard Teague: Regenerative Organic Practices 'Clean Up the Act of Agriculture,'" AgFunder Network Partners, June 21, 2019, https:// agfundernews.com/dr-richard-teague-regenerative-organic-practices -clean-up-the-act-of-agriculture.

22. "Dark Branch Farms Case Study," Understanding AG Case Studies, https:// understandingag.com/regenerative-successes/case-studies.

23. Paul Brown, "Gabe Brown: Keys to Building Healthy Soil," Brown's Ranch, video, 58:52, May 13, 2015, https://brownsranch.us/category/videos.

24. UnderstandingAG, "On the Brink of Foreclosure," Testimonials, https://understandingag.com/regenerative-successes/testimonials/.
25. Meeting Place Pastures, https://www.meetingplacepastures.com/.
26. Strailey, "Grass-Fed, Wild-Caught."
27. W. Richard Teague et al., "The Role of Ruminants in Reducing Agriculture's Carbon Footprint in North America," *Journal of Soil and Water Conservation* 7, no. 2 (2016): 6, https://doi.org/10.2489/jswc.71.2.156.
28. The Nutrition Source, "Protein," Harvard School of Public Health, https://www.hsph.harvard.edu/nutritionsource/what-should-you-eat/protein/.
29. Real Organic Project, Why We Exist, "Our mission is to grow people's understanding of foundational organic values and practices. . . . Crops grown in soil and livestock raised on pasture are fundamental to organic farming," https://www.realorganicproject.org/why-we-exist/ (accessed June 15, 2022).
30. European Commission, Directorate-General for Health and Food Safety, *Study on Intra European Union (Intra-EU) Animal Health Certification of Certain Live Animals: Final Report*, Publications Office, 2017, https://data.europa.eu/doi/10.2875/809481.

Chapter 2. The Empty Breadbasket

1. Ricardo Salvador, "Here's What Agriculture of the Future Looks Like: The Multiple Benefits of Regenerative Agriculture Quantified," *Union of Concerned Scientists* [blog], September 19, 2018, https://blog.ucsusa.org/ricardo-salvador/heres-what-agriculture-of-the-future-looks-like-the-multiple-benefits-of-regenerative-agriculture-quantified/.
2. M. Shahandeh, "Corn for Grain Production in the U.S.," *Statista*, January 14, 2022, https://www.statista.com/statistics/190871/corn-for-grain-production-in-the-us-since-2000/#statisticContainer.
3. Ramdas Kanissery et al., "Glyphosate: Its Environmental Persistence and Impact on Crop Health and Nutrition," *Plants* (Basel, Switzerland) 8, no. 11 (2019): 499, http://doi.com/10.3390/plants8110499.
4. R. Mesnage and M. N. Antoniou, "Contaminants," in *Encyclopedia of the Anthropocene*, ed. Dominick A. Dellasala and Michael I. Goldstein (2018), quoted in "Glyphosate: Contaminants," *ScienceDirect*, https://www.sciencedirect.com/topics/earth-and-planetary-sciences/glyphosate.
5. Magdalena Druille et al., "Arbuscular Mycorrhizal Fungi Are Directly and Indirectly Affected by Glyphosate Application," *Applied Soil Ecology* 72 (2013): 1, https://doi.org/10.1016/j.apsoil.2013.06.011.
6. Natural Resources Conservation Service (NRCS), "Soil Health Nuggets," USDA, https://www.nrcs.usda.gov/Internet/FSE_DOCUMENTS/stelprdb1101660.pdf.
7. Rita S. L. Veiga et al., "Can Arbuscular Mycorrhizal Fungi Reduce the Growth of Agricultural Weeds?" *PloS One* 6, no. 12 (2011): e27825, https://www.ncbi.nlm.nih.gov/pmc/articles/PMC3229497/.

8. W. Richard Teague et al., "The Role of Ruminants in Reducing Agriculture's Carbon Footprint in North America," *Journal of Soil and Water Conservation* 71, no. 2 (2016): 156–64, https://doi.org/10.2489/jswc.71.2.156.

9. Mike Amaranthus and Bruce Allyn, "Healthy Soil Microbes, Healthy People," *Atlantic*, June 11, 2013, https://www.theatlantic.com/health/archive /2013/06/healthy-soil-microbes-healthy-people/276710/.

10. Daan, "Why Ploughing Is Such a Bad Idea," Data Driven Investor, February 20, 2019, https://medium.datadriveninvestor.com/why-ploughing -is-such-a-bad-idea-62956c17967c.

11. Richard Bowen, "Ruminal Acidosis (Grain Overload)," Colorado State University, January 2020, http://www.vivo.colostate.edu/hbooks/pathphys /digestion/herbivores/acidosis.html.

12. Agricultural Marketing Service (AMS), statistics, https://www.ams.usda.gov.

13. USDA Office of Communications, "USDA Coexistence Fact Sheets: Corn," USDA, February 2015, https://www.usda.gov/sites/default/files/documents /coexistence-corn-factsheet.pdf; Ashley Broocks et al., "Corn as Cattle Feed vs. Human Food," Oklahoma Cooperative Extension Service, Oklahoma State University, March 2017, ANSI-3296, https://extension.okstate.edu/fact -sheets/print-publications/afs/corn-as-cattle-feed-vs-human-food-afs-3296.pdf.

14. Teague et al., "The Role of Ruminants": 157.

15. John Dobberstein, "No-Till Movement in U.S. Continues to Grow," *No-Till Farmer*, August 1, 2014, https://www.no-tillfarmer.com/articles/489-no-till -movement-in-us-continues-to-grow ; Frank Lessiter, "No Glyphosate Means Less No-Till," *No-Till Farmer*, August 3, 2020, https://www.no-tillfarmer.com /blogs/1-covering-no-till/post/9869-no-glyphosate-means-less-no-till.

16. Michael Pollan, "What's Eating America," *Smithsonian Magazine*, July 2006, https://www.smithsonianmag.com/history/whats-eating-america-121229356/.

17. Jeremy Woods et al., "Energy and the Food System," *Philosophical Transactions of the Royal Society of London B Biological Sciences 365*, no. 1554 (2010): 2991, https://doi.org/10.1098/rstb.2010.0172.

18. Teague et al., "The Role of Ruminants": 157.

19. For example, see Annie Lowrey, "Your Diet Is Cooking the Planet," *Atlantic*, April 6, 2021, https://www.theatlantic.com/health/archive/2021/04 /rules-eating-fight-climate-change/618515/. In concluding that all beef is bad for the planet, the author fails to examine the net climate benefit of regenerative grazing.

20. Teague et al., "The Role of Ruminants": 156–59.

21. Christine Jones, "Light Farming: Restoring Carbon, Organic Nitrogen and Biodiversity to Agricultural Soils," *Amazing Carbon* (2018): 6, http://amazingcarbon.com/JONES-LightFarmingFINAL(2018).pdf; Tracy Frisch, "Supporting the Soil Carbon Sponge," interview with Walter Jehne, *Eco Farming Daily*, https://www.ecofarmingdaily.com/

supporting-the-soil-carbon-sponge/; Elizabeth Ridlington et al., "Reaping What We Sow: How the Practices of Industrial Agriculture Put Our Health and Environment at Risk," *Frontier Group*, February 7, 2018, 16–17, https://frontiergroup.org /reports/fg/reaping-what-we-sow; Stephan van Vliet, Scott L. Kronberg, and Fredrick D. Provenza, "Plant-Based Meats, Human Health, and Climate Change," *Frontiers in Sustainable Food Systems*, October 6, 2020, https://doi.org/10.3389/fsufs.2020.00128; Paige L. Stanley et al., "Impacts of Soil Carbon Sequestration on Life Cycle Greenhouse Gas Emissions in Midwestern USA Beef Finishing Systems," *Agricultural Systems* 162 (May 2018): 250, https://doi.org/10.1016/j.agsy.2018.02.003.

22. Kurt Lawton, "Economics of Soil Loss," *Farm Progress*, March 13, 2017, https://www.farmprogress.com/soil-health/economics-soil-loss.

23. Sabrina Shankman, "What Is Nitrous Oxide and Why Is It a Climate Threat?" *Inside Climate News*, September 11, 2019, https://insideclimatenews.org /news/11092019/nitrous-oxide-climate-pollutant-explainer-greenhouse-gas -agriculture-livestock/.

24. Stanley et al., "Impacts of Soil Carbon Sequestration."

25. Rob Cook, "Ranking of States That Produce the Most Corn," *Beef 2 Live*, December 21, 2021, https://beef2live.com/story-states-produce-corn-0 -107129; Jan Dutton, "US Soybean Production by State: Top 11 Rankings," *Crop Prophet*, May 5, 2021, https://www.cropprophet.com/soybean -production-by-state-top-11/.

26. Fred Pearce, "Can the World Find Solutions to the Nitrogen Crisis?" *Yale Environment 360*, February 6, 2018, https://e360.yale.edu/features/can-the -world-find-solutions-to-the-nitrogen-pollution-crisis.

27. Pearce, "Can the World Find Solutions."

28. Richard Conniff, "The Nitrogen Problem and Why Global Warming Is Making It Worse," *Yale Environment 360*, August 7, 2017, https://e360.yale.edu /features/the-nitrogen-problem-why-global-warming-is-making-it-worse.

29. Conniff, "The Nitrogen Problem."

30. Evan A. Thaler, Isaac J. Larsen, and Qian Yu, "The Extent of Soil Loss across the US Corn Belt," *PNAS* 118, no. 8 (2021): e1922375118, http://doi.org /10.1073/pnas.1922375118.

31. Matt A. Sanderson, Leonard W. Jolley, and James P. Dobrowolski, "Pastureland and Hayland in the USA: Land Resources, Conservation Practices, and Ecosystem Services," in *Conservation Outcomes from Pastureland and Hayland Practices*, ed. C. Jerry Nelson (Lawrence, KS: Allen Press, 2012): 27–40, https://www.nrcs.usda.gov/Internet/FSE_DOCUMENTS/stelprdb 1080492.pdf.

32. Donald Stotts, "Carbon Sequestration a Positive Aspect of Beef Cattle Grazing Grasslands," Division of Agricultural Sciences and Natural Resources,

Oklahoma State University, http://www.dasnr.okstate.edu/Members
/donald-stotts-40okstate.edu/carbon-sequestration-a-positive-aspect-of
-beef-cattle-grazing-grasslands/.

Chapter 3. Cattle as Global Heroes

1. Bill McKibben, "The Only Way to Have a Cow," *Orion*, 2010, https://orion
magazine.org/article/the-only-way-to-have-a-cow/.
2. Llewellyn L. Manske, ed., "Biology of Defoliation by Grazing," in *Biogeo-
chemical Processes of Prairie Ecosystems*, 2nd ed. (Fargo: Dickenson Research
Extension Center, North Dakota State University, 2014), 6.
3. Manske, "Biology of Defoliation."
4. Lee Rinehart, "Building Healthy Pasture Soils," ATTRA Sustainable Agriculture (Butte MT: National Center for Appropriate Technology, October 2017,
IP546): 7, https://attra.ncat.org/product/building-healthy-pasture-soils/.
5. Rinehart, "Building Healthy Pasture Soils."
6. China is Uruguay's biggest market for all beef (Fermin Koop, "Uruguay Plans
to Boost Beef Production and Lessen Its Carbon Footprint," *Diálogo Chino*,
May 19, 2021, https://dialogochino.net/en/agriculture/uruguay
-strategy-boost-beef-china-lessen-footprint/), but the United States is the
biggest market for high-end cuts ("Uruguay's Meat Processing Industry: The
United States Is a Reliable Customer When Doing Business," US Embassy in
Uruguay, https://uy.usembassy.gov/uruguays-meat-processing-industry-the-
united-states-is-a-reliable-customer-when-doing-business/).
7. Richard Teague and Urs Kreuter, "Managing Grazing to Restore Soil Health,
Ecosystem Function, and Ecosystem Services," *Frontiers in Sustainable Food
Systems* 4, no. 106647 (September 29, 2020): https://doi.org/10.3389
/fsufs.2020.534187.
8. Christine Jones, "Soil Restoration: 5 Core Principles," *Eco Farming Daily*,
https://www.ecofarmingdaily.com/build-soil/soil-restoration-5-core
-principles/. This article originally appeared in the October 2017 issue of
Acres U.S.A. magazine.
9. W. Richard Teague et al., "The Role of Ruminants in Reducing Agriculture's
Carbon Footprint in North America," *Journal of Soil and Water Conservation*
71, no. 2 (2016): 156–64, https://doi.org/10.2489/jswc.71.2.156.
10. Steven I. Apfelbaum et al., "Vegetation, Water Infiltration, and Soil Carbon
Response to Adaptive Multi-Paddock and Conventional Grazing in Southeastern USA Ranches," *Journal of Environmental Management* 308 (April 15,
2022): 114576, https://doi.org/10.1016/j.jenvman.2022.114576.
11. Christine Jones, "Light Farming: Restoring Carbon, Organic Nitrogen, and
Biodiversity to Agricultural Soils," *Amazing Carbon*, http://amazingcarbon
.com/JONES-LightFarmingFINAL(2018).pdf: 6; Elaine Ingham in conversation with Ridge Shinn.

12. Jennifer Hayden, "Cattle Are Part of the Climate Solution: A Conversation with Rangeland Ecologist Richard Teague, PhD, Analyzing the Role That Adaptive Multi-Paddock Cattle Grazing Plays in Sequestering Carbon," Rodale Institute, August 28, 2020, https://rodaleinstitute.org/blog/cattle-are-part-of-the-climate-solution/.

13. W. Richard Teague, "Forages and Pastures Symposium: Cover Crops in Live-stock Production: Whole-System Approach: Managing Grazing to Restore Soil Health and Farm Livelihoods," *Journal of Animal Science* 96, no. 4 (2018): 1519, https://doi.org/10.1093/jas/skx060.

14. Jones, "Light Farming."

15. Tracy Frisch, "Interview: Supporting the Soil Carbon Sponge," *Eco Farming Daily*, https://www.ecofarmingdaily.com/supporting-the-soil-carbon-sponge/, originally published in *Acres U.S.A.* 49, no. 4 (April 2019).

16. Frisch, "Interview: Supporting the Soil Carbon Sponge."

17. Steven L. Dowhower et al., "Soil Greenhouse Gas Emissions as Impacted by Soil Moisture and Temperature under Continuous and Holistic Planned Grazing in Native Tallgrass Prairie," *Agriculture, Ecosystems and Environment* 287, no. 106647 (2020): https://doi.org/10.1016/j.agee.2019.106647.

18. Lan Liu et al., "Relationships between Plant Diversity and Soil Microbial Diversity Vary across Taxonomic Groups and Spatial Scales," *Ecosphere*, January 7, 2020, https://esajournals.onlinelibrary.wiley.com/doi/10.1002/ecs2.2999.

19. Chris Gill, "Desert Grasslands Restoration: *Manejo Holistico* in Chihuahua–Las Damas Ranch," *Pitchstone Waters*, June 15, 2015, https://pitchstonewaters.com/manejo-holistico-in-chihuahua-las-damas-ranch/.

20. "Las Damas Ranch Case Study," Understanding AG, https://understandingag.com/case_studies/las-damas-ranch-case-study/.

21. Judith D. Schwartz, "Hope for a Thirsty World," *Water in Plain Sight* (White River Junction, VT: Chelsea Green Publishing, 2019), adapted for the web at https://www.chelseagreen.com/product/water-in-plain-sight/.

22. Schwartz, "Hope for a Thirsty World."

23. Gabe Brown, "Dirt to Soil: Excerpt," *Resilience.org*, November 6, 2020, https://www.resilience.org/stories/2020-11-06/dirt-to-soil-excerpt/.

24. W. R. Teague et al., "The Role of Ruminants," *Journal of Soil and Water Conservation* 71, no. 2 (2016): 156–64, https://www.jswconline.org/content/71/2/156.

25. Ferris Jabr, "The Earth Is Just as Alive as You Are," *New York Times*, April 20, 2019, https://www.nytimes.com/2019/04/20/opinion/sunday/amazon-earth-rain-forest-environment.html.

26. Cheryl Anderson, "Rotational Grazing Is Green," *Progressive Farmer*, April 2016, https://static1.squarespace.com/static/58b5e62629687fdc87a1ad5b

/t/59160a16893fc0e34e149bde/1494616604828/Progressive+Farmer+%E2
%80%93+April+2016+-+Rotational+Grazing+Is+Green1.pdf.

27. Ronald S. Oremland and Charles W. Culbertson, "Importance of Methane-
Oxidizing Bacteria in the Methane Budget as Revealed by the Use of a
Specific Inhibitor," *Nature 356*, no. 6368 (1992): 421–23, https://doi.org
/10.1038/356421a0.

28. S. Tiwari et al., "Methanotrophs and CH4 Sink: Effect of Human Activity
and Ecological Perturbations," *Climate Change and Environmental Sustainability 3*, no. 1 (2015): 35–50.

29. Peter Bruce-Iri, "Methane Sources, Sinks, and Uncertainties," October 2021,
https://doi.org/10.13140/RG.2.2.28627.71201.

30. Paige Stanley et al., "Impacts of Soil Carbon Sequestration on Life Cycle
Greenhouse Gas Emissions in Midwestern USA Beef Finishing Systems,"
Agricultural Systems 162 (May 2018): 250, https://doi.org/10.1016/j.agsy
.2018.02.003.

31. "Climate Change: Seven Technology Solutions That Could Help Solve
Crisis," *Sky News*, October 12, 2021, https://news.sky.com/story/climate
-change-seven-technology-solutions-that-could-help-solve-crisis-12056397.

32. Stanley at al., "Impacts of Sequestration on Life Cycle Emissions."

33. Teague et al., "The Role of Ruminants."

34. Emily Payne, "Dr. Richard Teague: Regenerative Organic Practices
'Clean Up the Act of Agriculture,'" *AgFunder News*, June 21, 2019, https://
agfundernews.com/dr-richard-teague-regenerative-organic-practices-clean
-up-the-act-of-agriculture.html.

35. Megan B. Machmuller et al., "Emerging Land Use Practices Rapidly Increase
Soil Organic Matter," *Nature Communications 6*, no. 6995 (2015): https://doi
.org/10.1038/ncomms7995.

36. Samantha Mosier et al., "Adaptive Multi-Paddock Grazing Enhances
Soil Carbon and Nitrogen Stocks and Stabilization through Mineral Association in Southeastern U.S. Grazing Lands," *Journal of Environmental Management*
288, no. 112409 (2021): https://doi.org/10.1016/j.jenvman.2021.112409.

37. Hayden, "Cattle Are Part of the Climate Solution."

38. Dr. Rattan Lal, "Subcommittee on Clean Air, Climate Change, and Nuclear
Safety Hearing: Agricultural Sequestration to Address Climate Change
through Reducing Atmosphering Levels of Carbon Dioxide," U.S. Senate
Committee on Environment and Public Works, testimony, July 8, 2003,
https://www.epw.senate.gov/public/index.cfm/hearings?Id=DFC559DC
-802A-23AD-45F8-9E24C672229B&Statement_id=25034E6F-98C2-4928
-861D-E41DA0E9343B.

39. Allen Williams, "Can We Produce Grassfed Beef at Scale?" *Holistic Management International*, April 13, 2018, https://holisticmanagement.org/
featured-blog-posts/scaling-grassfed-beef-by-allen-williams/.

Chapter 4. The Roots of Health

1. David R. Montgomery and Anna Biklé, *What Your Food Ate* (New York: W. W. Norton Company, 2022).
2. Stephan van Vliet, Frederick D. Provenza, and Scott L. Kronberg, "Health-Promoting Phytonutrients Are Higher in Grass-Fed Meat and Milk," *Frontiers in Sustainable Food Systems* 4, no. 555426 (2021): https://doi.org/10.3389/fsufs.2020.555426.
3. Cynthia A. Daley et al., "A Review of Fatty Acid Profiles and Antioxidant Content in Grass-Fed and Grain-Fed Beef," *Nutrition Journal* 9, no. 10 (2010): https://doi.org/10.1186/1475-2891-9-10.
4. Frederick D. Provenza, Scott L. Kronberg, and Pablo Gregorini, "Is Grassfed Meat and Dairy Better for Human and Environmental Health?" *Frontiers in Nutrition* 6, no. 26 (2019): https://doi.org/10.3389/fnut.2019.00026.
5. Susanna C. Larsson and Nicola Orsini, "Red Meat and Processed Meat Consumption and All-Cause Mortality: A Meta-Analysis," *American Journal of Epidemiology* 179, no. 3 (2014): 282–89, https://doi.org/10.1093/aje/kwt261.
6. A. Wolk, "Potential Health Hazards of Eating Red Meat," *Journal of Internal Medicine*, September 6, 2016, https://onlinelibrary.wiley.com/doi/10.1111/joim.12543.
7. Renata Micha, Sarah K. Wallace, and Dariush Mozaffarian, "Red and Processed Meat Consumption and Risk of Incident Coronary Heart Disease, Stroke, and Diabetes Mellitus: A Systematic Review and Meta-Analysis," *Circulation* 121 (2010): 2271–83, https://pubmed.ncbi.nlm.nih.gov/20479151/.
8. Korin Miller, "Most Americans Are Not Meeting Cancer-Preventing Dietary Guidelines," *Verywell Health*, July 9, 2021, https://www.verywellhealth.com/cancer-preventing-dietary-guidelines-5191824.
9. Gina Kolata, "Eat Less Red Meat, Scientists Said. Now Some Believe That Was Bad Advice," *New York Times*, October 4, 2019, https://www.nytimes.com/2019/09/30/health/red-meat-heart-cancer.html.
10. Bradley C. Johnston et al., "Unprocessed Meat and Processed Meat Consumption: Dietary Guideline Recommendations from the Nutritional Recommendations (NutriRECS) Consortium," *Annals of Internal Medicine*, November 19, 2019, https://doi.org/10.7326/M19-1621.
11. Kolata, "Eat Less Red Meat, Scientists Said."
12. Mildred M. Haley, "Changing Consumer Demand for Meat: The US Example," Changing Structure of Global Food Consumption and Trade (Economic Research Service, USDA, WRS-01-1), https://www.ers.usda.gov/webdocs/outlooks/40303/14976_wrs011g_1_.pdf?v=9429.1.
13. Amy Norton, "Metabolic Syndrome Continues to Climb in the U.S.," *Reuters Health*, October 15, 2010, https://www.reuters.com/article/us-metabolic-syndrome-idUSTRE69E5FL20101015.

14. Nicolette Hahn Niman, *Defending Beef: The Ecological and Nutritional Case for Meat*, 2nd ed. (White River Junction, VT: Chelsea Green Publishing, 2021).

15. Ben Webster, "Women Who Eat Too Little Meat and Dairy Put Their Health at Risk, Says Scientist," *The Times*, January 12, 2022, https://www.thetimes .co.uk/article/women-who-eat-little-meat-and-dairy-put-their-health-at-risk -says-scientist-ch2dz0z58.

16. Cleveland Clinic, "Do I Need to Worry about Eating 'Complete' Proteins?" *Health Essentials*, March 12, 2019, https://health.clevelandclinic.org/do-i -need-to-worry-about-eating-complete-proteins/.

17. "Dietary Protein Quality Evaluation in Human Nutrition: Report of an FAO Expert Consultation," FAO Food and Nutrition Paper 92, 2011, https:// www.fao.org/ag/humannutrition/35978-02317b979a686a57aa4593304 ffc17f06.pdf.

18. Stephan van Vliet, Frederick D. Provenza, and Scott L. Kronberg, "Plant-Based Meats, Human Health, and Climate Change," *Frontiers in Sustainable Food Systems* 4, no. 128 (2020): https://doi.org/10.3389/fsufs.2020.00128.

19. Joseph Opoku Gakpo, "FAO Predicts Global Shortage of Protein-Rich Foods," *Alliance for Science*, July 7, 2020, https://allianceforscience.cornell .edu/blog/2020/07/fao-predicts-global-shortage-of-protein-rich-foods/.

20. Richard Hurrell and Ines Egli, "Iron Bioavailability and Dietary Reference Values," *American Journal of Clinical Nutrition* 91, no. 5 (2010): 1461S–67S, https://doi.org/10.3945/ajcn.2010.28674F.

21. Van Vliet et al., "Plant-Based Meats."

22. J. Efren Ramirez Bribiesca et al., "Supplementing Selenium and Zinc Nanoparticles in Ruminants for Improving Their Bioavailability in Meat," in *Nutrient Delivery*, vol. 5 of *Nanotechnology in the Agri-Food Industry*, ed. Alexandru Grumezescu (Elsevier, 2017): 713–47, https://doi.org/10.1016 /B978-0-12-804304-2.00019-6.

23. M. Foster and S. Samman, "Chapter 3: Vegetarian Diets across the Lifecycle: Impact on Zinc Intake and Status," *Advances in Food Nutrition Research* 74 (2015): 93–131, https://doi.org/10.1016/bs.afnr.2014.11.003.

24. Van Vliet et al., "Plant-Based Meats."

25. Van Vliet et al., "Plant-Based Meats."

26. Agriculture Research Service, "Beef, Ground, Unspecified Fat Content, Cooked," Food Data Central Search Results, USDA, April 1, 2019, https:// fdc.nal.usda.gov/fdc-app.html#/food-details/172161/nutrients.

27. Van Vliet et al., "Health-Promoting Phytonutrients."

28. Robert K. Lyons and Richard V. Machen, "Interpreting Grazing Behavior," RangeDetect Series (AgriLife Communications and Marketing, Texas A&M University, 2012), https://animalscience.tamu.edu/wp-content/uploads /sites/14/2012/04/beef-interpreting-grazing-behavior.pdf.

29. Van Vliet et al., "Health-Promoting Phytonutrients."

30. The Nutrition Source, "Riboflavin—Vitamin B2," T. H. Chan, Harvard School of Public Health, https://www.hsph.harvard.edu/nutritionsource/riboflavin-vitamin-b2/; Office of Dietary Supplements, National Institutes of Health, "Thiamin," National Institutes of Health, US Department of Health and Human Services, https://ods.od.nih.gov/factsheets/Thiamin-HealthProfessional/.

31. Van Vliet et al., "Health-Promoting Phytonutrients."

32. Van Vliet et al., "Health-Promoting Phytonutrients."

33. Harvard Health Publishing, "The Truth about Fats: The Good, the Bad, and the In-Between: Avoid the Trans Fats, Limit the Saturated Fats, and Replace with Essential Polyunsaturated Fats," Harvard Medical School, December 11, 2019, https://www.health.harvard.edu/staying-healthy/the-truth-about-fats-bad-and-good.

34. Philip C. Calder, "Functional Roles of Fatty Acids and Their Effects on Human Health," *Journal of Parenteral and Enteral Nutrition* 39, no. 1 Suppl (2015): 18S–32S, https://doi.org/10.1177/0148607115595980.

35. Elizabeth Quin, "Converting Fat to Energy During Exercise," *Verywell Fit: Sports Nutrition*, October 12, 2021, https://www.verywellfit.com/sports-nutrition-how-fat-provides-energy-for-exercise-3120664.

36. World Health Organization, "Obesity and Overweight," Fact Sheet, June 9, 2021, https://www.who.int/news-room/fact-sheets/detail/obesity-and-overweight.

37. R. K. Miller, "Chemical and Physical Characteristics of Meat: Palatability," in *Encyclopedia of Meat Sciences*, 2nd ed., 3 vols., ed. Michael Dikeman and Carrick Devine (Elsevier, 2014), https://www.sciencedirect.com/topics/veterinary-science-and-veterinary-medicine/intramuscular-fat.

38. Jonathan A. Campbell, "Understanding Beef Carcass Yields and Losses During Processing," Penn State Extension, College of Agricultural Sciences, Penn State University, August 4, 2016.

39. K. R. Matthews, "Saturated Fat Reduction in Butchered Meat," in *Reducing Saturated Fats in Foods*, ed. Geoff Talbot (Philadelphia: Woodhead Publishing, 2011), https://www.sciencedirect.com/topics/veterinary-science-and-veterinary-medicine/intramuscular-fat.

40. Provenza et al., "Is Grassfed Meat and Dairy Better?"

41. Daley et al., "A Review of Fatty Acid Profiles."

42. American College of Cardiology, "Cardiovascular Disease Burden, Deaths Are Rising around the World," December 9, 2020, https://www.acc.org/about-acc/press-releases/2020/12/09/18/30/cvd-burden-and-deaths-rising-around-the-world.

43. Stephanie Venn-Watson, Richard Lumpkin, and Edward A. Dennis, "Efficacy of Dietary Odd-Chain Saturated Fatty Acid Pentadecanoic Acid Parallels Broad Associated Health Benefits in Humans: Could It Be Essential?"

Scientific Reports 10, no. 8161 (2020), https://doi.org/10.1038/s41598-020 -64960-y.

44. Venn-Watson et al., "Efficacy of Dietary Odd-Chain Saturated Fatty Acid Pentadecanoic Acid."

45. Daley et al., "A Review of Fatty Acid Profiles."

46. Amit Sachdeva et al., "Lipid Levels in Patients Hospitalized with Coronary Artery Disease: An Analysis of 136,905 Hospitalizations in Get With The Guidelines," *American Heart Journal* 157, no. 1 (2009): 111–17.e2, https://doi .org/10.1016/j.ahj.2008.08.010.

47. Alice H. Lichtenstein, "2021 Dietary Guidance to Improve Cardiovascular Health: A Scientific Statement from the American Heart Association," *Circulation* 144, no. 23 (2021): e472–e487, https://doi.org/10.1161/CIR .0000000000001031; US Department of Agriculture and US Department of Health and Human Services, Dietary Guidelines for Americans, 2020–2025, 9th ed. (December 2020), https://www.dietaryguidelines.gov/sites/default /files/2021-03/Dietary_Guidelines_for_Americans-2020-2025.pdf.

48. Kathy Trieu et al., "Biomarkers of Dairy Intake, Incident Cardiovascular Disease, and All-Cause Mortality: A Cohort Study, Systematic Review, and Meta-Analysis," *PLOS Medicine* 18, no. 9 (September 21, 2021): e1003763, https://doi.org/10.137/journal.pmed.1003763.

49. Anamaria Balic et al., "Omega-3 versus Omega-6 Polyunsaturated Fatty Acids in the Prevention and Treatment of Inflammatory Skin Diseases," *International Journal of Molecular Sciences* 21, no. 3 (2020): 741, https://doi .org/10.3390/ijms21030741.

50. Balic et al., "Omega-3 versus Omega-6."

51. Provenza et al., "Is Grassfed Meat and Dairy Better?"

52. Daley et al., "A Review of Fatty Acid Profiles."

53. A. J. McAfee et al., "Red Meat from Animals Offered a Grass Diet Increases Plasma and Platelet n-3 PUFA in Healthy Consumers," *British Journal of Nutrition* 105, no. 1 (2010): 80–89, https://doi.org/10.1017/S0007114510003090.

54. Danielle Swanson, Robert Block, and Shaker A. Mousa, "Omega-3 Fatty Acids EPA and DHA: Health Benefits throughout Life," *Advances in Nutrition* 3, no. 1 (1): 1–7, https://doi.org/10.3945/an.111.000893; Adda Bjarnadottir, "DHA (Docosahexaenoic Acid): A Detailed Review," *Healthline*, updated May 28, 2019, https://www.healthline.com/nutrition/dha-docosa hexaenoic-acid.

55. Daley et al., "A Review of Fatty Acid Profiles."

56. Marko Dachev et al., "The Effects of Conjugated Linoleic Acids on Cancer," *Processes* 9, no. 3 (2021): 454, https://doi.org/10.3390/pr9030454.

57. Daley et al., "A Review of Fatty Acid Profiles."

58. Laura M. den Hartigh, "Conjugated Linoleic Acid Effects on Cancer, Obesity, and Atherosclerosis: A Review of Pre-Clinical and Human Trials with

Current Perspectives," *Nutrients* 11, no. 2 (2019): 370, https://doi.org /10.3390/nu11020370.

59. K. W. J. Wahle and S. D. Heys, "Cell Signal Mechanisms, Conjugated Linoleic Acids (CLAs) and Anti-Tumorigenesis," *Prostaglandins, Leukotrienes & Essential Fatty Acids* 67, nos. 2–3 (2002): 183–86, https://doi.org/10.1054 /plef.2002.0416.

60. Dachev et al., "The Effects of Conjugated Linoleic Acids": 454.

61. Susanna C. Larsson, Leif Bergkvist, and Alicja Wolk, "High-Fat Dairy Food and Conjugated Linoleic Acid Intakes in Relation to Colorectal Cancer Incidence in the Swedish Mammography Cohort," *American Journal of Clinical Nutrition* 82, no. 4 (2005): 894–900, https://doi.org/10.1093/ajcn/82.4.894.

62. Dachev et al., "The Effects of Conjugated Linoleic Acids": 10.

63. Renee Cheung and Paul McMahon et al., *Back to Grass: The Market Potential for U.S. Grassfed Beef* (Stone Barns Center for Food & Agriculture, Armonia LLC, Bonterra Partners, SLM Partners, 2017): 14, https://www.stonebarns center.org/wp-content/uploads/2017/10/Grassfed_Full_v2.pdf.

64. Dated documents from Clemson University in the files of Ridge Shinn; available upon request.

65. Julius Ruechel, "Marbling in Grass-Fed Beef," Grass Fed Solutions, accessed November 7, 2021, https://www.grass-fed-solutions.com/marbling.html.

66. Tracy A. McCrorie et al., "Human Health Effects of Conjugated Linoleic Acid from Milk and Supplements," *Nutrition Research Reviews* 24, no. 2 (2011): 206–7, https://doi.org/10.1017/S0954422411000114.

67. Dachev et al., "The Effects of Conjugated Linoleic Acids": 11.

68. Daley et al., "A Review of Fatty Acid Profiles."

69. Daley et al., "A Review of Fatty Acid Profiles."

70. Van Vliet et al., "Plant-Based Meats."

71. Jerome A. Paulson et al., "Nontherapeutic Use of Antimicrobial Agents in Animal Agriculture: Implications for Pediatrics," *Pediatrics* 136, no. 6 (2015): e1670-e1677, https://doi.org/10.1542/peds.2015-3630.

72. Paulson et al., "Nontherapeutic Use of Antimicrobial Agents."

73. Centers for Disease Control and Prevention, "2019 AR Threats Report," Antibiotic/Antimicrobial Resistance (AR/AMR Resistance), Atlanta, GA, available at www.cdc.gov/drugresistance/threat-report-2013/ (accessed October 27, 2021).

74. "Leftovers for Livestock: A Legal Guide for Using Food Scraps as Animal Feed," Harvard Food Law and Policy Clinic and Food Recovery Project at the University of Arkansas School of Law, August 2016, https://further withfood.org/wp-content/uploads/2017/07/Leftovers-for-Livestock _A-Legal-Guide_August-2016.pdf.

75. F. M. Pate and W. F. Brown, "Improving the Value of Hydrolyzed Feather Meal as a Protein Source in a Molasses-Based Liquid Supplement Fed to

Cattle," University of Florida, Range Cattle Research and Education Center, October 1994.

76. Paulson et al., "Nontherapeutic Use of Antimicrobial Agents."

77. Mayo Clinic, "*E. coli*," https://www.mayoclinic.org/diseases-conditions /e-coli/symptoms-causes/syc-20372058.

78. Rick Stock and Robert Britton, "Beef Cattle Handbook: Acidosis," BCH-3500, 2014 (Extension Beef Cattle Resource Committee, adapted from the Cattle Producer's Library, CL624), https://www.iowabeefcenter.org/bch /Acidosis.pdf.

79. H. Roger Segelken, "Simple Change in Cattle Diets Could Cut *E. coli* Infection," *Cornell Chronicle*, September 8, 1998, https://news.cornell.edu /stories/1998/09/simple-change-cattle-diets-could-cut-e-coli-infection.

80. "USDA Confirms Sixth Case of Mad Cow Disease in Past 15 Years," *Food Safety News*, August 30, 2018, https://www.foodsafetynews.com/2018/08 /usda-confirms-sixth-case-of-mad-cow-disease-in-past-15-years/.

81. David Thomas, "The Mineral Depletion of Foods Available to US as a Nation (1940–2002)—A Review of the 6th Edition of McCance and Widdowson," *Nutrition and Health* 19 (2007): 21–55, https://doi.org/10.1177/026010 600701900205.

82. Donald R. Davis, "Declining Fruit and Vegetable Nutrient Composition: What Is the Evidence?" *HortScience* 44, no. 1 (2009): 15–19, https://doi.org /10.21273/HORTSCI.44.1.15.

83. Tracy Frisch, "Supporting the Soil Carbon Sponge," *Acres U.S.A.*, 2019, https://www.ecofarmingdaily.com/supporting-the-soil-carbon-sponge/.

84. W. Richard Teague et al., "The Role of Ruminants in Reducing Agriculture's Carbon Footprint in North America," *Journal of Soil and Water Conservation* 71, no. 2 (2016): 158, https://doi.org/10.2489/jswc.71.2.156.

85. Tracy Frisch, "SOS: Save Our Soils. Dr. Christine Jones Explains the Life-Giving Link between Carbon and Healthy Topsoil," *Acres U.S.A.* 45, no. 3 (March 2015).

86. Frisch, "SOS: Save Our Soils."

87. Provenza et al., "Is Grassfed Meat and Dairy Better?"

Chapter 5. Animal Welfare

1. Michael Pollan, "Power Steer," *New York Times Magazine*, March 31, 2002.

2. American Humane, "Five Freedoms: The Gold Standard of Animal Welfare," October 17, 2016, https://www.americanhumane.org/blog/five-freedoms -the-gold-standard-of-animal-welfare/.

3. Victoria Burnett, "Avocados Imperil Monarch Butterflies' Winter Home in Mexico," *New York Times*, November 18, 2016, https://www.nytimes .com/2016/11/18/world/americas/ambition-of-avocado-imperils-monarch -butterflies-winter-home.html.

4. M. Ancrenaz et al., "Bornean Orangutan: *Pongo pygmaeus*," IUCN Red List of Threatened Species, 2016, https://doi.org/10.2305/IUCN.UK.2016-1.RLTS.T17975A17966347.en.

5. Stephan van Vliet, Scott L. Kronberg, and Frederick D. Provenza, "Plant-Based Meats, Human Health, and Climate Change," *Frontiers of Sustainable Food Systems* 4, no. 128 (2020), https://doi.org/10.3389/fsufs.2020.00128.

6. Van Vliet et al., "Plant-Based Meats."

7. M. Shahbandeh, "Total Slaughtered Cattle in the US 2000–2019," *Statista*, January 28, 2022, https://www.statista.com/statistics/194357/total-cattle-slaughter-in-the-us-since-2000/.

8. NRDC, "Feedlot Operations: Why It Matters Where Your Grain-Finished Beef Was Produced," November 13, 2014, https://www.nrdc.org/resources/feedlot-operations-why-it-matters.

9. Microbe Wiki, "Bovine Rumen," updated July 7, 2011, https://microbewiki.kenyon.edu/index.php/Bovine_Rumen.

10. Matt Maier, "2019 Thousand Hills Lifetime Grazed™ 100% Grass Fed Beef Program Regenerative Protocol Requirements," Thousand Hills Cattle Company, October 10, 2018, https://thousandhillslifetimegrazed.com/wp-content/uploads/2019/06/2019-THLG-Regenerative-Protocol.pdf.

11. Tara L. Felix, "Use of Beta-Agonists in Cattle Feed," Penn State Extension, September 7, 2017, https://extension.psu.edu/use-of-beta-agonists-in-cattle-feed.

12. "Cattle and Beef: Sector at a Glance," USDA Economic Research Service, last updated April 12, 2022, https://www.ers.usda.gov/topics/animal-products/cattle-beef/sector-at-a-glance/.

13. Jerome A. Paulson et al., "Nontherapeutic Use of Antimicrobial Agents in Animal Agriculture: Implications for Pediatrics," *Pediatrics* 136, no. 6 (2015): e1670-e1677, https://doi.org/10.1542/peds.2015-3630.

14. Joaquin Hernandez et al., "Ruminal Acidosis in Feedlot: From Aetiology to Prevention," *Scientific World Journal* 2014, no. 702572 (2014): https://www.hindawi.com/journals/tswj/2014/702572/.

15. T. G. Nagaraja and M. M. Chengappa, "Liver Abscesses in Feedlot Cattle: A Review," *Journal of Animal Sciences* 76, no. 1 (January 1998): 287–98, https://doi.org/10.2527/1998.761287x.

16. South Dakota State University, "Antibiotics in Cattle," Beef2Live, December 7, 2021, https://www.beef2live.com/story-antibiotics-cattle-88-106554.

17. Michael Priestly, "Worries Raised Over Zilmax Audit," The Beef Site, November 6, 2013, https://www.thebeefsite.com/news/44290/worries-raised-over-zilmax-audit/.

18. Temple Grandin, "Cattle and Pigs [That] Are Easy to Move and Handle Will Have Less Preslaughter Stress," *Foods* 10, no. 11 (November 2021): 2583, https://doi.org/10.3390/foods10112583.

19. Julianne Johnston, "Tyson Notifies Producers of Plan to Ban Cattle Fed Beta Agonists," *Farm Journal* blog, *Drovers*, August 8, 2013, https://www.drovers.com/news/tyson-notifies-producers-plan-ban-cattle-fed-beta-agonists.

20. Temple Grandin, *Guide to Working with Farm Animals* (North Adams, MA: Storey Publishing, 2018): 161.

21. Bruno I. Cappellozza and Rodrigo S. Marques, "Effects of Pre-Slaughter Stress on Meat Characteristics and Consumer Experience," in *Meat and Nutrition*, ed. Chhabi Lal Ranabhat (ebook chapter: IntechOpen, 2021), https://www.intechopen.com/chapters/75636.

22. Temple Grandin, "The Effect of Stress on Livestock and Meat Quality Prior to and During Slaughter," *International Journal for the Study of Animal Problems* 1, no. 5 (1980): 313–37, https://www.wellbeingintlstudies repository.org/cgi/viewcontent.cgi?article=1019&context=acwp_faafp.

23. "Government Oversight and Inspection," AnimalHandling.org, https://animalhandling.org/consumers/government_oversight_inspection.

24. Michael Holtz, "6 Months Inside One of America's Most Dangerous Industries," *Atlantic*, July–August 2021, https://www.theatlantic.com/magazine/archive/2021/07/meatpacking-plant-dodge-city/619011/.

25. Grandin, *Guide to Working with Farm Animals*: 169.

26. "Life and Death from the Soil Food Web on Up," *Regenetarianism* (formerly *L.A. Chefs Column*), https://lachefnet.wordpress.com/2021/06/14/life-and-death-from-the-soil-food-web-on-up/.

27. Nicolette Hahn Niman, *Defending Beef: The Ecological and Nutritional Case for Meat*, 2nd ed. (White River Junction, VT: Chelsea Green Publishing, 2021).

Chapter 6. Achieving Wholesale Benefits

1. Mary Berry and Debbie Barker, "Renewing a Vision for Rural Prosperity," *Civil Eats*, August 15, 2018, https://civileats.com/2018/08/15/renewing-a-vision-for-rural-prosperity-in-wendell-berry-country/.

Chapter 7. Turning a Profit

1. Steven I. Apfelbaum et al., "Vegetation, Water Infiltration, and Soil Carbon Response to Adaptive Multi-Paddock and Conventional Grazing in Southeastern USA Ranches," *Journal of Environmental Management* 308 (April 15, 2022): 114576, https://doi.org/10.1016/j.jenvman.2022.114576.

2. Agronomic Crops Network, "The Haney Test for Soil Health," *C.O.R.N. Newsletter* 7 (2019), College of Food, Agricultural, and Environmental Sciences, Ohio State University Extension, https://agcrops.osu.edu/newsletter/corn-newsletter/2019-07/haney-test-soil-health.

3. Noted by Doug Gunnick, founder, Minnesota Intensive Graziers Groups, at a conference presentation in Hardwick, MA, 2004.

4. John Kempf, "The Challenges of Managing Nutrition with Brix Readings," blog entry, *Advancing Eco Agriculture*, August 21, 2020, https://johnkempf .com/the-challenges-of-managing-nutrition-with-brix-readings/.

5. Lisa Abend, "How Cows (Grass-Fed Only) Could Save the Planet," *TIME*, January 25, 2010, https://content.time.com/time/subscriber/article /0,33009,1953692,00.html.

6. Emma Neuberger, "'It Crushes Me': Dairy Farmers Struggle to Survive Trump's Trade Wars and Declining Milk Demand," CNBC, January 4, 2020, https://www.cnbc.com/2020/01/04/us-dairy-farmers-battle-extinction -trump-trade-wars-lower-milk-prices.html.

Chapter 8. Cattle into Beef

1. George McRobie, "The Community's Role in Appropriate Technology," 2nd Annual E. F. Schumacher Lecture, given at Cathedral of St. John the Divine, New York, October 1982, Schumacher Center for a New Economics, https://centerforneweconomics.org/publications/the-communitys -role-in-appropriate-technology/.

2. House Agricultural Committee, "Chairman David Scott Opening Statement at Hearing 'An Examination of Price Discrepancies, Transparency, and Alleged Unfair Practices in Cattle Markets,'" press release, July 27, 2021, https:// agriculture.house.gov/news/documentsingle.aspx?DocumentID=2491.

3. The White House Briefing Room, "Fact Sheet: The Biden-Harris Action Plan for a Fairer, More Competitive, and More Resilient Meat and Poultry Supply Chain," January 3, 2022, https://www.whitehouse.gov/briefing-room /statements-releases/2022/01/03/fact-sheet-the-biden-harris-action-plan-for -a-fairer-more-competitive-and-more-resilient-meat-and-poultry-supply-chain/.

4. Andrew Lisa, "History of America's Meat Processing Industry," *Stacker*, August 18, 2020, https://stacker.com/stories/4402/history-americas-meat -processing-industry.

5. White House Briefing Room, "Fact Sheet."

6. Tom Polansek, "Explainer: How Four Big Companies Control the U.S. Beef Industry," *Reuters*, June 17, 2021, https://www.reuters.com/business/how -four-big-companies-control-us-beef-industry-2021-06-17/.

7. NBCA, "Fast Facts: State of U.S. Cattle Ranching," May 21, 2018, https:// www.neogen.com/neocenter/blog/fast-facts-state-of-u-s-cattle-ranching/.

8. Michael Holtz, "6 Months Inside One of America's Most Dangerous Industries: What I Learned on the Line at a Dodge City Slaughterhouse," *Atlantic*, June 14, 2021, https://www.theatlantic.com/magazine/archive /2021/07/meatpacking-plant-dodge-city/619011/.

9. Jeremiah Moss, "Meatpacking Before & After," *Jeremiah's Vanishing New York* blog, August 28, 2013, http://vanishingnewyork.blogspot.com/2013/08 /meatpacking-before-after.html.

10. Alexandra Thorn, Michael J. Baker, and Christian J. Peters, "Estimating Biological Capacity for Grass-Finished Ruminant Meat Production in New England and New York," *Agricultural Systems* 189 (April 2021): 102958, https://doi.org/10.1016/j.agsy.2020.102958.

11. Peter Thomas Ricci, "Paradigm SHIFT," *Meatingplace Desktop*, January 2022, http://library.meatingplace.com/publication/frame.php?i=734718&p=&pn=&ver=html5&view=articleBrowser&article_id=4193556.

12. Andrea Shalal, "Meat Packers' Profit Margins Jumped 300% During Pandemic—White House Economics Team," *Reuters*, December 10, 2021, https://www.reuters.com/business/meat-packers-profit-margins-jumped-300-during-pandemic-white-house-economics-2021-12-10/.

13. Polansek, "Explainer: How Four Big Companies."

14. Phil McCausland, "The Price of Meat Is Going Up. Ranchers and Corporations Are Split on Why," *A NoBull News Service*, October 2, 2021, https://news.mikecallicrate.com/nbc-news-the-price-of-meat-is-going-up-ranchers-and-corporations-are-split-on-why/.

15. White House Briefing Room, "Fact Sheet."

Chapter 9. Public Awareness, Public Policies

1. Wendell Berry, "The Pleasures of Eating," *What Are People For* (San Francisco: North Point Press, 1990), available at https://emergencemagazine.org/essay/the-pleasures-of-eating/.

2. "Farm Subsidy Primer," Environmental Working Group (EWG), https://farm.ewg.org/subsidyprimer.php.

3. Anne Schechinger, "Under Trump, Farm Subsidies Soared and the Rich Got Richer," Environmental Working Group (EWG), February 24, 2021, https://www.ewg.org/interactive-maps/2021-farm-subsidies-ballooned-under-trump/.

4. Adam Andrzejewski, "Mapping the U.S. Farm Subsidy $1M Club," *Forbes*, August 14, 2018, https://www.forbes.com/sites/adamandrzejewski/2018/08/14/mapping-the-u-s-farm-subsidy-1-million-club/?sh=6df3381d3efc.

5. "Fair Prices for Farmers: A Fair Price = A Living Wage," National Family Farm Coalition (NFFC), https://nffc.net/what-we-do/fair-prices-for-farmers/.

6. Tara O'Neill Hayes and Katerina Kerska, "PRIMER: Agriculture Subsidies and Their Influence on the Composition of U.S. Food Supply and Consumption," American Action Forum, November 3, 2021, https://www.americanactionforum.org/research/primer-agriculture-subsidies-and-their-influence-on-the-composition-of-u-s-food-supply-and-consumption/.

7. Heather Cox Richardson, "Corn in the USA," *The Historical Society: A Blog Devoted to History for the Academy and Beyond*, posted February 5, 2010, http://histsociety.blogspot.com/2010/11/corn-and-united-states.html.

8. Union of Concerned Scientists, "Turning Soils into Sponges: How Farmers Can Fight Floods and Droughts," August 7, 2017, https://www.ucsusa.org /resources/turning-soils-sponges.

9. Ronnie Cummins, *Grassroots Rising: A Call to Action on Climate, Farming, Food, and a Green New Deal* (White River Junction, VT: Chelsea Green Publishing, 2020).

10. Agricultural Marketing Service, "Country of Origin Labeling (COOL)," US Department of Agriculture, https://www.ams.usda.gov/rules-regulations/cool.

11. The White House Briefing Room, "Fact Sheet: The Biden-Harris Action Plan for a Fairer, More Competitive, and More Resilient Meat and Poultry Supply Chain," January 3, 2022, https://www.whitehouse.gov/briefing-room /statements-releases/2022/01/03/fact-sheet-the-biden-harris-action-plan-for -a-fairer-more-competitive-and-more-resilient-meat-and-poultry-supply-chain/.

12. Ryan Nebeker and Jerusha Klemperer, "The FoodPrint of Fake Meat," *Food-Print*, updated February, 7, 2022, https://foodprint.org/reports/the-foodprint -of-fake-meat/#main-content.

13. Stephan van Vliet, Scott L. Kronberg, and Frederick D. Provenza, "Plant-Based Meats, Human Health, and Climate Change," *Frontiers in Sustainable Food Systems* 4 (October 6, 2020): 128, https://www.frontiersin.org/articles /10.3389/fsufs.2020.00128/full.

14. Van Vliet et al.,, "Plant-Based Meats."

15. Nebeker et al.,, "The FoodPrint of Fake Meat."

16. Kayla Butera, "From Lab to Table: How the Kaplan Lab Is Pioneering Cellular Agriculture," *The Tufts Daily*, January 21, 2022, https://tuftsdaily.com /features/2022/01/21/from-lab-to-table-how-the-kaplan-lab-is-pioneering -cellular-agriculture/.

17. Butera, "From Lab to Table."

18. Butera, "From Lab to Table."

19. "The Romance versus the Reality of Cultured Stem Cell Proteins," *Regenetarianism* (formerly *L.A. Chef's Column*), https://lachefnet.wordpress.com/2021/02/19 /the-romance-versus-the-reality-of-cultured-stem-cell-proteins/.

20. Nebeker et al.,, "The FoodPrint of Fake Meat."

21. Butera, "From Lab to Table."

22. Butera, "From Lab to Table."

23. "Climate Change: Seven Technology Solutions That Could Help Solve Crisis," *Sky News*, October 12, 2021, https://news.sky.com/story/climate -change-seven-technology-solutions-that-could-help-solve-crisis-12056397.

24. "Climate Change: Seven Technology Solutions."

25. Ken Caldeira, "Chapter 6: Ocean Storage: Introduction and Background," in *IPCC Special Report on Carbon Dioxide Capture and Storage,* eds. Brad De Young and Fortunat Joos (2005): 279, https://www.ipcc.ch/site/assets/uploads /2018/03/srccs_chapter6-1.pdf.

26. Caldeira, "Ocean Storage."

27. Market Intel, "Too Many to Count: Factors Driving Fertilizer Prices Higher and Higher," American Farm Bureau Federation, December 13, 2021, https://www.fb.org/market-intel/too-many-to-count-factors-driving-fertilizer-prices-higher-and-higher.

28. Beth Brelje, "Farmers Paying Triple for Fertilizer: We're Not the Ones Raising Food Costs," *The Epoch Times*, March 25, 2022. https://www.theepochtimes.com/farmers-paying-triple-for-fertilizer-were-not-the-ones-raising-food-costs_4354883.html.

29. Rushan Chai et al., "Greenhouse Gas Emissions from Synthetic Nitrogen Manufacture and Fertilization for Main Upland Crops in China," *Carbon Balance and Management*, https://cbmjournal.biomedcentral.com/articles/10.1186/s13021-019-0133-9.

30. Jerome A. Paulson et al., "Nontherapeutic Use of Antimicrobial Agents in Animal Agriculture: Implications for Pediatrics," *Pediatrics* 136, no. 6 (2015): e1670–e1677, https://doi.org/10.1542/peds.2015-3630.

31. "Northern Gulf of Mexico Hypoxic Zone, EPA, June 3, 2021, https://www.epa.gov/ms-htf/northern-gulf-mexico-hypoxic-zone.

32. "Health Rankings," American Public Health Association, 2021, https://www.apha.org/topics-and-issues/health-rankings.

33. Joe Fassler, "Regenerative Agriculture Needs a Reckoning," *The Counter*, May 3, 2021, https://thecounter.org/regenerative-agriculture-racial-equity-climate-change-carbon-farming-environmental-issues/.

34. Megan Horst, "New Research Explores the Ongoing Impact of Racism on the U.S. Farming Landscape," *Civil Eats*, January 25, 2019, https://civileats.com/2019/01/25/new-research-explores-the-ongoing-impact-of-racism-on-the-u-s-farming-landscape/.

35. Digital Public Library of America, https://dp.la/primary-source-sets/the-homestead-acts.

36. Horst, "New Research Explores the Ongoing Impact of Racism."

37. Mary McIntyre, "This Rancher Has Her Boots on the Ground," *Deseret News*, March 8, 2020, https://www.deseret.com/2022/3/8/22966198/this-rancher-has-her-boots-on-the-ground-dx-ranch-dx-beef-south-dakota.

38. Naomi Starkman, "On the Rural Immigrant Experience: 'We Come with a Culture, Our Own History, and We're Here to Help,'" *Civil Eats*, April 14, 2022, https://civileats.com/2022/04/14/on-the-rural-immigrant-experience-we-come-with-a-culture-our-own-history-and-were-here-to-help/.

39. Tom Maurer, "Parity Pricing Is the Solution [Opinion]," *Lancaster Farming*, March 25, 2022, https://www.lancasterfarming.com/parity-pricing-is-the-solution-opinion/article_4edde1c4-5ede-55d9-850f-d04a6da98fa0.html.

40. The *Kiss the Ground* website has a page devoted to the upcoming Farm Bill at https://kisstheground.com/regenerate-america/.

Index

About the Authors

Ridge Shinn is the founding CEO of Grazier LLC, aka Big Picture Beef, a 100% grass-fed beef company partnering with farmers throughout the Northeast United States. Early in his career he became interested in heritage breeds of livestock and cofounded the group now known as The Livestock Conservancy. He was also the founding director of the New England Livestock Alliance, which helped farmers  find markets for their meat. In addition to managing his Devon herd in central Massachusetts, Ridge has consulted all over North America, in New Zealand, England, Uruguay, and Argentina, and for the Lakota of the Cheyenne River Indian Reservation. His work has been recognized in *Smithsonian*, the *Atlantic*, the *New York Times*, and *TIME* magazine, which dubbed him a "carbon cowboy."

Warren Johnson

Lynne Pledger is a writer and environmental advocate. She has worked on public policy issues such as waste reduction, climate change, and energy in affiliation with nonprofit organizations, including Clean Water Action, Sierra Club, and Upstream, and has been a guest lecturer on sustainability topics at the University of Massachusetts Amherst,  Smith College, Lesley University, and the Harvard School of Public Health. She has also worked for decades with Ridge Shinn to preserve heritage livestock breeds and increase regenerative grazing in the Northeast United States. In addition, she homeschooled her two children and one grandchild. Now living in western Massachusetts, she is writing a book of poetry.

Jon Morris

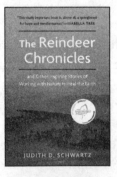